Fundamentals of

POLYMER
ENGINEERING

Fundamentals of

POLYMER ENGINEERING

Arie Ram

TECHNION-ISRAEL INSTITUTE OF TECHNOLOGY
HAIFA, ISRAEL

PLENUM PRESS
New York • London • Moscow

Library of Congress Cataloging-in-Publication Data

Ram, Arie.
    Fundamentals of polymer engineering / Arie Ram.
        p.   cm.
    Includes bibliographical references and index.
    ISBN 0-306-45726-1
    1. Polymers.   I. Title.
TP1087.R36   1997
668.9--dc21   97-41616
                                                                CIP

ISBN 0-306-45726-1

© 1997 Plenum Press, New York
A Division of Plenum Publishing Corporation
233 Spring Street, New York, N. Y. 10013

http://www.plenum.com

Book Designed by Reuven Solomon

Printed in the United States of America

To my beloved wife
SARA who patiently
did the hard work
on the hard disk

Preface

W<small>E ALL ARE SURROUNDED</small> by plastic materials and cannot imagine modern life and utilities without the synthetic polymers. And yet, how many of us can distinguish between polyethylene and PVC? After all, most people name any polymer as "Nylon." Is there any distinction between polymers and plastics?

This introductory textbook tries to answer these questions and many others. It endeavors to provide the basic information required in modern life about the best utilization of new materials in the plastics era; the chemical sources of synthetic polymers, and the processes in which small "simple" molecules are converted to giant macromolecules, namely, high polymers; and the understanding of the role of these unique structures, their behavior and performance, their mechanical and thermal properties, flow and deformation.

As we are mainly interested in the final product, the processing of plastics, through shaping and forming, presents a significant challenge to polymer engineering. All this is broadly discussed, ending with modern issues like composites, ecology and future prediction, followed by up-to-date information and data about old as well as novel high performance polymers.

The text is particularly targeted towards senior students of science and engineering (chemical, material, mechanical and others) who may use it as the first window to the world of polymers. At the same time many professionals who are involved in the resin or plastics industry may prefer this approach without elaborate math or overloading.

This book is mainly based on thirty-five years of teaching polymer engineering at the Technion, Haifa, Israel, and several universities in the United States.

Acknowledgments

I WOULD LIKE TO acknowledge the Departments of Chemical Engineering at the Technion, Haifa, and UCLA, Los Angeles, for providing the facilities and environment for writing this textbook.

Many thanks are due to Sidney and Raymond Solomon, of the Solomon Press, and to Mary Russell who carried out the professional work of editing, design, layout and publishing, with great care and tolerance.

Last but not the least, I wish to thank my numerous students, from whom I learned the most.

—A.R.

Contents

List of Photos

List of Tables

List of Figures

List of Symbols

SYMBOL	DEFINITION
a	exponent in eq. 3-5
A	area
AO	antioxidant
b	width
B	second virial coefficient (eqs. 3-11, 3-12)
C	concentration
CAD	computer aided design
CAM	computer aided manufacture
Cp	specific heat
d	thickness; diameter; density
D	diameter; diffusivity
De	Deborah number
DP	degree of polymerization (number of mers)
DSC	differential scanning calorimetry
E	energy of activation; Young's modulus; electrical potential
ESC	environomental stress cracking

f	initiator efficiency (eq. 2-42)
f_i	mole fraction of component i in feed mixture (for copolymerization)
F	force
F_i	mole fraction of component i in copolymer chain
G	shear modulus
G'	storage modulus
G''	loss modulus
G^*	complex modulus
GPC	gel permeation chromatography
h	depth of screw channel (extruder)
H	optical constant (eq. 3-12)
I	initiator concentration; current (electrical)
J	compliance
k	kinetic constant; heat conduction coefficient
k_d	initiator decomposition rate constant
k_i	rate constant of initiation
k_p	rate constant of propagation; rate constant of polymerization
k_t	rate constant of termination
$k_{t,c}$	rate constant of termination via combination
$k_{t,d}$	rate constant of termination via disproportionation
k_{tr}	rate of constant for transfer
ℓ	length
L	length
LCP	liquid crystal polymer
M	monomer concentration
M_o	monomer molecular weight
M^*	monomer radical concentration
$\overline{M}_n$	number-average molecular weight
$\overline{M}_v$	viscosity-average molecular weight
$\overline{M}_w$	weight-average molecular weight
$\overline{M}_z$	z average molecular weight

$\overline{M}_{z+1}$	$z+1$ average molecular weight
MFI	melt flow index
MSW	municipal solid waste
MW	molecular weight
MWD	molecular weight distribution
n	number of moles; number of repeating units (mers); exponent in eq. 3-15 (Avrami); exponent in power law (eq. 4-3)
N	number of functional groups (in condensation), speed of rotation
N_o	initial number of functional groups
P	polymer concentration; pressure; load; permeability; degree of conversion
ppb	parts per billion
ppm	parts per million
q_d	drag flow rate
q_ℓ	leakage flow rate
q_p	pressure flow rate
Q	flow rate (volumetric); electrical charge
r_i	rate of initiation
r_p	rate of propagation; rate of polymerization
r_t	rate of termination
R	gas constant; end-to-end distance; electrical resistivity
R^*	primary radical concentration
RIM	reactive injection concentration
S	tensile strength; solubility
S_B	tensile strength at break
S_y	tensile strength at yield
t	time; thickness; pitch (extruder)
t_c	cooling time (injection molding)
t_f	filling time (injection molding)
t_n	"dead" time (injection molding)
t_s	cycle time (injection molding)

T	temperature
T_c	temperature of crystallization
T_d	heat distortion temperature
T_f	temperature at opening of the mold
T_g	glass transition temperature
T_i	temperature of injection
T_m	melting temperature
T_s	surface temperature (mold)
u	velocity
U	velocity
UHMW	ultra high MW
UV	ultraviolet
UVA	ultraviolet absorber
V	velocity; volume; specific volume; elution volume
V_a	specific volume of amorphous phase
V_c	specific volume of crystalline phase
w_i	weight fraction of species i
W	weight
W_i	weight fraction
x_i	mole fraction of species i
X	mole ratio in the feed (copolymerization)
Y	mole ratio in the copolymer chain

Greek Symbols

SYMBOL	DEFINITION
α	thermal expansion coefficient; heat diffusivity
γ	strain; surface energy
$\dot{\gamma}$	rate of shear (velocity gradient)
δ	shift angle; depth of deformation
ϵ	dielectric constant
η	viscosity

η_o	low-shear viscosity (Newtonian)
η_{Tg}	viscosity at T_g
θ	dimensionless temperature, shift angle
λ	relaxation time, retardation time
μ	viscosity (Newtonian); micron
μ_e	effective viscosity
μ_r	relative viscosity
μ_s	viscosity of solvent
μ_{sp}	specific viscosity
$[\mu]$	intrinsic viscosity
ν	Poisson ratio
π	osmotic pressure
Π	power
ρ	resistivity (electrical); density
τ	turbidity; shear stress
ϕ	angle of inclination (extruder screw)
ϕ	shift angle
ω	frequency

1

Introduction to the World of Polymers

WHAT IS THE MEANING of *polymers*?

This word stems from Greek: Poly = many, mers = particles. So this term describes a molecule composed of many identical parts, called mers. The large molecule is therefore termed: macromolecule. What is left for us, is the practical definition of the term "many". The minimum considers hundreds of mers. However, there are no significant mechanical properties below about 30 mers, while the useful average reaches 200–2,000 mers. If one wants to speak about molecular weights (which is usually done in chemistry) a broad range between 5,000 up to 2×10^6 may be representative, while in some cases it may reach 10^7.

The sheer existence of such a broad range of molecular weights indicates that we deal with long molecules that have no fixed or standard length (or weight). This presents a most comprehensive difference between macromolecules and the smaller molecules that are characterized by a single and fixed molecular weight (e.g. water = 18). Prior to delving deep into the world of polymers, it is essential also to explain the common term "plastics". The name itself actually, describes the stage of processing the polymer while it is plastic or soft—enabling smooth flow and shaping. On the other hand, plastics (or the scientific term, plastomers) refer to the major group of polymers that in combination with various additives leads to materials of construction.

In addition to plastomers, polymers are also used as elastomers (rubber-like), textiles, coatings and adhesives. Many polymers may appear in various utility groups, determined by the final desired composition. Because plastics use is dominated by polymers, many people do not care about the difference between those two terms, which are similar but by no means identical.

While the synthetic polymers are, for good reasons, of major interest in this book one should acknowledge the historical role of natural polymers since the beginning of mankind: in food (protein, starch and others); in clothing (wool, cotton, silk); and in various other uses (cellulose for writing and natural resins for ornaments). Later (in the 18th century) the resin exuded from rubber trees (Hevea Braziliensis) turned into a very useful product in transportation (wheels and tires), in industry (conveyor belts) and in general uses, including toys. Even today, many natural polymers are still in use,

TABLE 1-1
Historical Development of Production of Synthetic Polymers

Polymer	Year	Polymer	Year
Celluloid	1870	Polypropylene	1959
Phenolics	1909	Polyacetal (Delrin™)	1959
Alkyds	1926	Chlorinated polyether	1959
Aniline-formaldehyde	1926	(Penton™)	
Cellulose acetate	1927	Phenoxy	1962
Urea formaldehyde	1928	Ionomers	1964
Polymethyl methacrylate	1931	Polyphenylene oxide (Noryl™)	1964
Ethyl cellulose	1931	Parylenes	1965
Polyvinyl chloride (PVC)	1936	Polybutylene	1965
Polyvinyl acetate	1938	Poly (4-methyl pentene) TPX	1965
Polyvinylbutyral	1938	Polybenzimidazole	1965
Polystyrene	1938	Polysulfones	1965
Polyethylene (LD)	1939	Polyethylene sulfide	1966
Polyamides (Nylon 6-6)	1939	Trogamid-T	1966
Melamines	1939	Polyaryl ether	1968
Polyvinylidene chloride	1939	Polyphenylene sulfide	1968
Polyesters (unsaturated)	1942	Polybenzimidazo benzophenan-	1970
Silicones	1943	throline (ladder)	
Polytetrafluoro ethylene (Teflon™)	1943	Polybutylene terephthalate (PBT)	1971
Polyethylene terephthalate (PET)	1945	Aromatic polyamides (Kevlar™)	1972
		Polyamide-imide (PAI)	1973
ABS	1946	Aromatic polyester	1974
Epoxy	1947	Polyarylate (Ardel™)	1978
Polyurethanes	1953	Polyether-ether ketone (PEEK)	1982
Polyethylene (HD)	1954	Polyetherimide (Ultem™)	1982
Polycarbonate	1958	Thermotropic polyesters (LCP)	1984

reaching highly developed processing methods. However, modern industry is mainly based on raw material that is suitable for mass production. Hence the vitality of the synthetic polymers.

In intermediate stages (the 19th century) polymers made by the modification of natural resins have appeared, the most prominent ones being the cellulose derivatives. Celluloid (obtained by nitration of cellulose) represents the first semi-synthetic polymer. It became useful after compounding with a plasticizer (mainly camphor) to reduce its brittleness. Many cellulose derivatives are still currently in use (as plastomers, textiles or coatings) but the major development in the 20th century is definitely attributed to many families of synthetic polymers—the era of polymers.

Table 1-1 presents by historical year, the beginning of commercial production of the most useful polymers. While many useful polymers appeared in the 1930s, the process of introducing new polymers continued during the 1950s, 1960s and so on. It apparently takes about 10 years for a newborn polymer to reach maturity. During this period it has to undergo infant development and compete with other polymers or nonpolymeric materials. The major issue has become the tremendous cost of commercialization of a new polymer, because the period of research and development (including marketing) is long and costly. In spite of all this, novel polymers appear every year, as long as they demonstrate uniqueness. However, there appears an increasing trend towards polyblends (mixtures of existing polymers). There is a distinction between homogeneous polymers (homopolymers), consisting of identical mers, and heterogeneous ones made of a random combination of two (or more) mers, namely, the copolymers. The latter may involve various compositions differing in the type, concentration and order of the distinct mers in the macromolecule.

In conclusion, a wide array of polymers is already in use, so that a systematic presentation of their chemical structure, as well as the relationship between structure and behavior at the final stage, is essential.

PROBLEMS

1. Describe 3 products made of plastics. What are the advantages over other materials? State other options.

2. Describe 3 natural polymers and the scope of applications. What synthetic polymers may replace them and what are their advantages?

3. Describe 5 thermoplastic polymers and 5 thermosetting polymers.

What are the principal differences between the two groups? Give details on the structure of the mer in each case.

4. Can you convert thermoplastic polymers to thermosetting polymers? Thermosets to thermoplastics? Describe hybrids.

5. Find the output of polymer production in the recent two years in the US. Give details of production of the major polymer families and state the changes for each of them.

6. Which polymers are products from the coal industry and which are produced by the petrochemical industry?

CHAPTER

2

The Chemistry of Polymers

2.1 INTRODUCTION

*I*N THIS CHAPTER we deal with a very basic concept—from where does everything start? We begin with the raw materials for the polymer industry, the so-called monomers, and explain how they are produced. The process for polymer synthesis—polymerization—should be well understood regarding both the mechanism as well as the industrial technology. We avoid going into too many details but still point out the basic principles.

Polymerization reactions involve careful control of purity of monomers, ratio of reactants, special additives, temperature and pressure, separation and recovery. Each family of polymers represents a wide range of variables. They appear in a large number of grades, differing in molecular weight, distribution, degree of crystallinity, size of particles and types of special additives. The polymer leaves the reactor as a powder or as pellets (often, after passing through extrusion and granulation). Some stabilizers are directly added in the polymerization or granulation stage. The polymer is then stored in big silos for homogenation. This process is mostly carried out at special compounding units, which are responsible for exact mixing and composition.

2.2 THE PETROCHEMICAL INDUSTRY

The petrochemical industry, that branch of the chemical industry which is based on the exploitation of the crude-oil distillation products, has turned out to be the leading industry in chemistry, wherein most monomers and polymers are produced. It was preceded by the coal industry, developed mainly in Germany and Britain during the 19th century. By decomposition of coal at high temperature (an anaerobic process called cracking) products like acetylene, methanol, or phenol are derived. These chemicals serve as the primary source for an extended array of polymers.

Acetylene is derived from carbide, the latter obtained by a reaction between lime and coke (the major solid fraction obtained from coal pyrolysis). The basic chemical reactions are as follows:

$$CaO + 3C \rightarrow CaC_2 + CO \tag{2-1}$$

$$CaC_2 + 2H_2O \rightarrow Ca(OH)_2 + HC\equiv CH \tag{2-2}$$
$$\text{(acetylene)}$$

Just another step leads to the manufacture of one of the oldest monomers — vinylchloride — along with other vinyl derivatives (including acrylonitrile).

$$CH\equiv CH + HCl \rightarrow CH_2=CHCl \tag{2-3}$$
$$\text{(vinylchloride)}$$

Ethylene, which serves as a source of many other monomers, can also be synthesized by the hydrogenation of acetylene.

$$CH\equiv CH + H_2 \rightarrow CH_2=CH_2 \tag{2-4}$$

Methanol is obtained in another process, by oxidizing coke with steam to form a mixture of carbon monoxide and hydrogen, as follows:

$$C + H_2O \rightarrow CO + H_2 \tag{2-5}$$

Next steps will lead to the formation of methanol via a catalytic reaction:

$$CO + 2H_2 \rightarrow CH_3OH \tag{2-6}$$

Methanol may be further oxidized to formaldehyde:

$$CH_3OH + \tfrac{1}{2}O_2 \rightarrow HCOH \tag{2-7}$$
$$\text{(formaldehyde)}$$

A whole family of polymers is derived from formaldehyde: polyacetal (polymethylene oxide), phenol-formaldehyde, urea-formaldehyde and melamine-formaldehyde. It is interesting to note that the other components (phenol, urea and melamine) are also products of coal pyrolysis.

Like many other aromatics, phenol is produced from another fraction of the cracking of coal, coal-tar. These aromatics serve as the basis for a large list of other polymers, like nylons, epoxides and polycarbonates, in addition to polystyrene and other styrenic derivatives (ABS, SB rubber) and polyesters.

The petrochemical industry (based on crude oil and partly on natural gas) essentially replaced the coal industry as the major source of monomers. This industry was developed in Europe during the 1950s, but started in the United States as early as 1920.

The lowest homolog in the paraffin family, methane (CH_4), (derived from crude oil but frequently found in natural gas), serves as the basis for the manufacture of methanol:

$$CH_4 + H_2O \rightarrow CO + 3H_2 \tag{2-8}$$

$$CO + 2H_2 \rightarrow CH_3OH \tag{2-9}$$

Alternatively, methane can be converted to acetylene, and through it to all kinds of monomers.

$$2CH_4 \rightarrow C_2H_2 + 3H_2 \tag{2-10}$$

Currently, by cracking the light fraction naphtha (with a boiling point between gasoline and kerosene), the unsaturated gases that serve the primary monomers — ethylene, propylene and butylene — as well as aromatics (including phenol) are obtained. From these, many monomers are derived. By the 1960s, 90% of all organic chemicals were derived from oil, and this trend continued growing. Only 3% to 5% of crude oil is used as chemicals, while the major portion is utilized as fuel. The world forecast for production of petrochemicals during 1996 was:

Ethylene	65 million tons
Propylene	40 million tons
Butadiene	7 million tons
Benzene	38 million tons
Xylenes	28 million tons

Let us describe the routes to some selected monomers, produced by the petrochemical industry.

Monomers Derived from Ethylene

(VCM) $CH_2{=}CHCl$
   (vinyl chloride)

$$CH_2=CH_2 + 2HCl + \tfrac{1}{2}O_2 \rightarrow CH_2Cl-CH_2Cl \rightarrow \qquad (2\text{-}11)$$
$$CH_2=CHCl + HCl$$

VCM can also be manufactured by an alternative reaction:

$$CH_2=CH_2 + Cl_2 \rightarrow CH_2Cl-CH_2Cl \rightarrow CH_2=CHCl \qquad (2\text{-}12)$$

Styrene can be synthesized by reacting ethylene with benzene (the latter present in the aromatic fraction of the oil cracking process – benzene, toluene and xylene).

$$2CH_2=CH_2 + 2C_6H_6 \rightarrow C_6H_5-C_2H_5 + \underset{\underset{C_6H_5}{|}}{CH_2=CH} \qquad (2\text{-}13)$$

(styrene)

Styrene serves as the monomer for the well-known polymer – polystyrene. It also serves as the source of many copolymers, that is polymers made from two monomers at varying compositions, such as SAN = styrene-acrylonitrile; SBR = styrene-butadiene rubber (the major synthetic rubber); SBS = styrene-butadiene-styrene (a modern family of thermoplastic rubbers which are not cross-linked); and the well-known terpolymer ABS which is based on 3 monomers – acrylonitrile-butadiene-styrene.

Vinyl acetate (a monomer frequently used in adhesives and coatings) may also be synthesized by reacting ethylene with acetic acid:

$$H_2C=CH_2 + CH_3COOH \rightarrow \underset{\underset{O}{\overset{\|}{}}}{CH_2=CH-O-C-CH_3} \qquad (2\text{-}14)$$

Monomers Derived from Propylene

Acrylonitrile is obtained by reacting propylene with ammonia:

$$\underset{\underset{CH_3}{|}}{CH_2=CH} + NH_3 + 3/2O_2 \rightarrow \underset{\underset{CN}{|}}{H_2C=CH} + 3H_2O \qquad (2\text{-}15)$$

Another important monomer, methylmethacrylate (acrylic), is obtained by reacting propylene with carbon monoxide, oxygen and methanol:

$$H_2C=CH + CO + \tfrac{1}{2}O_2 + CH_3OH \rightarrow H_2C=\underset{\underset{COOCH_3}{|}}{\overset{\overset{CH_3}{|}}{C}} + H_2O \qquad (2\text{-}16)$$

$$(MMA)$$

Propylene also serves as the source of the epoxy polymer, via epichlorohydrin:

$$H_2C=\underset{\underset{CH_3}{|}}{CH} \overset{\overset{Cl_2}{\downarrow}}{\rightarrow} H_2C=CH-CH_2Cl \rightarrow H_3C-\underset{\underset{OH}{|}}{CH}-CH_2Cl \rightarrow$$

$$\underset{\underset{O}{\diagdown \diagup}}{H_2C-CH-CH_2Cl} \qquad (2\text{-}17)$$

$$(epichlorohydrin)$$

Incidentally, by a series of oxidation reactions on propylene, a very useful polyol (glycerol) can be derived. The latter serves mainly in polyfunctional thermosets such as alkyds that undergo cross-linking.

$$H_2C=\underset{\underset{CH_3}{|}}{CH} + O_2 \rightarrow H_2C=\underset{\underset{H}{|}}{C}-\underset{\underset{H}{|}}{C}=O \rightarrow \qquad (2\text{-}18)$$

$$H_2C=CH-CH_2OH + H_2O_2 \rightarrow \underset{\underset{CH_2OH}{|}}{\underset{CHOH}{\overset{CH_2OH}{|}}}$$

$$(glycerol)$$

Phenol and Other Aromatics

Phenol is mainly derived from benzene, but can also be obtained from propylene. It serves as the basis for several important monomers, in addition to its direct use as a reactive monomer via condensation with formaldehyde. This first man-made polymer, phenol-formaldehyde, appeared in 1901 and bears the trademark Bakelite®, after its inventor Bakeland. By reacting phenol with acetone, bis-phenol A is obtained. This serves as the basis for the manufacture of epoxy and polycarbonate.

$$2\,C_6H_5OH + H_3C-\overset{\overset{\textstyle O}{\|}}{C}-CH_3 \rightarrow HO-C_6H_4-\overset{\overset{\textstyle CH_3}{|}}{\underset{\underset{\textstyle CH_3}{|}}{C}}-C_6H_4-OH + H_2O$$

(2-19)

(bis-phenol A)

Urea is obtained by the reaction between ammonia (NH_3) and carbon dioxide (CO_2):

$$2NH_3 + CO_2 \rightarrow H_2N-\overset{\overset{\textstyle}{}}{\underset{\underset{\textstyle O}{\|}}{C}}-NH_2 + H_2O$$

(2-20)

(urea)

As stated previously, condensation between urea and formaldehyde leads to the formation of the polymer urea-formaldehyde. A third member in this series is the melamine, which will also react with formaldehyde leading to the polymer melamine-formaldehyde. Melamine may also be derived by thermal decomposition of urea.

$$6\,H_2N-CO-NH_2 \rightarrow C_3N_6H_6 + 6\,NH_3 + 3\,CO_2$$

(2-21)

(melamine)

Both monomers for the production of Nylon 6–6 (one of the earliest polyamides) may be synthesized from benzene: adipic acid and hexamethylene diamine.

Phthalic acids (iso, tera and ortho) are manufactured in the petrochemical industry from the three isomers of xylene. These acids and the corresponding anhydrides are the major basis for polyesters and alkyds.

Needless to say, there are many monomers and we illustrated some of the most useful.

In conclusion, the petrochemical industry currently supplies most monomers, while the coal-cracking industry remains an alternative option.

2.3 MONOMERS: BUILDING BLOCKS FOR POLYMER MANUFACTURE

What is a monomer? Any material that is able to be polymerized (that is, can form macromolecules) possesses a unique chemical structure, which is termed **polyfunctionality**. The simplest case — representing a functionality of two — appears as the covalent double bond, resulting from electron sharing. The paraffins (like methane and its homologs) are unable to polymerize — due to lack of a double bond—hence they are not considered to be monomers. Ethylene, on the other hand, represents the simplest and most commonly used monomer, essentially due to the existence of an aliphatic double bond that starts the polymeric chain.

$$CH_2=CH_2 \rightarrow -CH_2-CH_2- \qquad (2\text{-}22)$$

$$\text{monomer n } CH_2=CH_2 \rightarrow (CH_2-CH_2)_n \text{ (polymer)} \qquad (2\text{-}23)$$

The various mechanisms of polymerization will be described later. If the monomer contains two double bonds (as in butadiene and other dienes) the final polymer can remain unsaturated, as each unit (mer) includes one double bond.

$$nCH_2=CH-CH=CH_2 \rightarrow (CH_2-CH=CH- CH_2)_n \qquad (2\text{-}24)$$

Ethylene represents a whole family of monomers, called vinyls

$$\begin{array}{c} \text{H} \quad \text{H} \\ | \quad\ | \\ \text{C}=\text{C} \\ | \quad\ | \\ \text{H} \quad \text{X} \end{array}$$

in which the substituent X is hydrogen (H). However, X may also represent other elements or radicals such as chlorine or other halogens; CH_3 group in propylene; phenyl (benzene ring) in styrene; an esteric group (acrylics); the CN group as in acrylonitrile; and many others. Sometimes two (or even more) of the hydrogen atoms are replaced. A list of monomers that have the potential to be polymerized through an addition reaction is shown in Table 2-1.

Another mode of bifunctionality exists in cyclic monomers, where opening of the ring is the starting stage of polymerization. The most common monomer in this group is ethylene oxide:

$$\begin{array}{c} CH_2-CH_2 \rightarrow CH_2-CH_2O \rightarrow -(CH_2-CH_2-O)_n \\ \diagdown \quad \diagup \\ O \end{array} \qquad (2\text{-}25)$$

The ring in order to be activated should consist of a number of sides differing from six. In both cases (double bonds or ring opening), the monomers are directly added into macromolecules.

Functionality is even more apparent in monomers that polymerize through a condensation reaction. Let us consider a condensation reaction of monofunctional groups, like classical esterification:

$$CH_3COOH + HOC_2H_5 \rightarrow H_3CCOOC_2H_5 + H_2O \qquad (2\text{-}26)$$

TABLE 2-1
Monomers Polymerized by Addition

Code and Name	Monomer
PE: Polyethylene	$CH_2=CH_2$
PP: Polypropylene	$HC=CH_2$ $\|$ CH_3
PIB: Polyisobutylene	CH_3 $\|$ $C=CH_2$ $\|$ CH_3
PMMA: Polymethylmethacrylate	CH_3 $\|$ $C=CH_2$ $\|$ $COOCH_3$
PAN: Polyacrylonitrile	$CH_2=CH$ $\|$ CN
PTFE: Polytetrafluoroethylene	$F_2C=CF_2$
POM: Polyoxymethylene (Polyacetal)	$CH_2=O$
PB: Polybutadiene	$H_2C=CH-CH=CH_2$
PVAL: Polyvinylalcohol	$HC=CH_2$ $\|$ OH
PB: Polybutene-1 (Polybutylene)	$CH=CH_2$ $\|$ $CH_2 CH_3$
TPX: Polymethylpentene	$CH=CH_2$ $\|$ CH_2 $\|$ CH $\diagup \diagdown$ $H_3C \;\; CH_3$
PVC: Polyvinylchloride	$ClCH=CH_2$
PVDC: Polyvinylidenchloride	$Cl_2 C=CH_2$
PVAC: Polyvinylacetate	$CH=CH_2$ $\|$ $OCOCH_3$
PS: Polystyrene	$CH=CH_2$ $\|$ $C_6 H_5$

It is apparent that two reactants—(an organic acid and an alcohol) consisting of single reactive (functional) groups of opposing nature (OH = hydroxyl and COOH = carboxyl)—"neutralize" each other. The resultant product is an ester, and one molecule of water is eliminated. This reaction also resembles the neutralization of an acid by a base (forming a salt) in inorganic chemistry; however, there is no sequence or chain reaction, so there is no polymer. If we want to build a polymer chain, at least bifunctional reactants are essential. In the case of the ester, a dicarboxylic acid and glycol are reacted—again via condensation.

$$HOOC(CH_2)_4COOH + HOCH_2-CH_2OH \rightarrow \qquad (2\text{-}27)$$
$$HOOC(CH_2)_4COOCH_2CH_2OH$$

The uniqueness of this ester is the retention of two opposing groups at the ends, which actually enables more and more reactions, thus building a large ester molecule—a polyester.

$$n[HOOC(CH_2)_4COOH] + n[HOH_2C-CH_2OH] \rightarrow \qquad (2\text{-}28)$$
$$HO[OC(CH_2)_4COOCH_2CH_2O]_nH + (2n - 1)H_2O$$

In conclusion, the existence of at least two functional groups is a necessary (yet not sufficient) requirement for a small molecule to act as a monomer. The repeating unit—the mer—is not necessarily identical to the monomer. Such an identity exists only in addition polymerization, wherein the monomer molecules link up without any elimination. In the case of polyesterification (as well as with other similar condensation reactions) elimination of small molecules is involved (in most cases water). That is the source of the name—condensation polymerization—a small molecule is evolved or condensed. The term "stepwise polymerization" is more commonly used. This will be explained later.

Another example of polymerization by condensation is the formation of polyamide (in our case Nylon 6-6) by reacting a di-acid with a diamine.

$$n[H_2N(CH_2)_6NH_2 + n[HOOC(CH_2)_4COOH] \rightarrow \qquad (2\text{-}29)$$
$$H[HN(CH_2)_6NHCO(CH_2)_4CO]_nOH + (2n - 1)H_2O$$

It is also possible to polymerize by condensation a single monomer with two opposing groups at its ends. In the case of a polyamide, Nylon 6 can be polymerized from the monomer $H_2N(CH_2)_5COOH$

$$n[H_2N(CH_2)_5COOH] \rightarrow H[HN(CH_2)_5CO]_nOH + (n - 1)H_2O \qquad (2\text{-}30)$$

In practice, the cyclic caprolactam anhydride serves as the common monomer.

$$n[\text{HN (CH}_2)_5 \text{ CO]} \rightarrow [\text{HN(CH}_2)_5 \text{ CO]}_n$$

(2-31)

The rate of polymerization is dependent on the structure of the monomer and the substituent groups, the right use of catalysts and initiators, and the appropriate conditions. All this will be thoroughly discussed in the next section that deals with the mechanism of polymerization.

Table 2-2 describes some monomers that polymerize by stepwise reactions (mostly condensation).

2.4 POLYMERIZATION—THE MECHANISM OF THE PROCESS

There are many mechanisms describing the process of polymerization (two of which were already discussed—addition and condensation). These terms describe only the chemical process, not the major engineering aspect—the rate of the reaction (kinetics). Therefore, it is beneficial to differentiate between two major polymerization routes—chain polymerization and stepwise polymerization.

Chain polymerization is mainly represented by the addition reaction and is characterized as follows:

1. Long chains appear at the early stage.
2. Monomers are added to long chains, and steadily disappear during the process.
3. There is no elimination of small molecules.
4. Quite high molecular weights may be obtained (10^5 to 2×10^6).
5. Extension of polymerization only increases conversion not the chain length.

On the other hand, the kinetics of stepwise polymerization differ completely, according to the following:

1. At the early stages dimers, trimers, and tetramers are formed, by reacting pairs of opposing functional groups, so that the growth develops in stages.
2. Monomers disappear in the early stage.
3. There is (usually) an elimination of a small molecule.
4. Molecular weights are low to medium (below 50,000 normally) .
5. Extension of the polymerization increases both conversion and molecular weights.

TABLE 2-2
Monomers Reacting by Stepwise Polymerization

Polymer	Monomers
PES: Polyester (PET: Polyethyleneterephthalate) Typical group	$HOCH_2-CH_2OH$ $HOOCC_6H_4COOH$ $-CO-O$
PA: Polyamide (Nylon 6-6) Typical group	$H_2N(CH_2)_6NH_2$ $HOOC(CH_2)_4COOH$ $-NH-CO-$
PUR: Polyurethane Typical group:	$HO(CH_2)_2OH$ $OCN(CH_2)_nNCO$ $-NH-CO-O-$
(Note: here there is no condensation.) PC: Polycarbonate Bis-phenol A Phosgene	 $HO-C_6H_4-\overset{\displaystyle CH_3}{\underset{\displaystyle CH_3}{\overset{\textstyle \vert}{\underset{\textstyle \vert}{C}}}}-C_6H_4-OH$ $Cl-\overset{\displaystyle O}{\overset{\Vert}{C}}-Cl$
PF: Phenol-formaldehyde	C_6H_5OH $HCHO$
UF: Urea-formaldehyde	$H_2N-CO-NH_2$ $HCHO$
MF: Melamine-formaldehyde	$H_2N-C\overset{\displaystyle N}{\underset{\displaystyle N}{}}\ \ C-NH_2$ ... $HCHO$

Stepwise polymerization includes most condensation processes, but also some reactions involving ring opening, and even addition reactions (typified by the formation of polyurethane from a di-isocyanate and a di-alcohol). The viscosity serves as a sensitive indication of the growth of the polymer. Therefore, in stepwise polymerization the viscosity of the system increases slowly and reaches high levels only toward the end of the reaction. In chain polymerization, high viscosity material appears at the initial stages, indicating a mixture (or solution) of long chains with yet unpolymerized monomers.

Chain (Addition) Polymerization

With this background, let us explain the mechanism of chain (addition) polymerization. If we write the electronic structure of the main group, vinyl:

$$
\begin{array}{ccc}
\underset{\underset{\displaystyle H}{|}}{\overset{\overset{\displaystyle H}{|}}{C}}=\underset{\underset{\displaystyle X}{|}}{\overset{\overset{\displaystyle H}{|}}{C}}
& \xrightarrow[\leftarrow]{}
\underset{H\ \ X}{\overset{H\ \ H}{C::C}}
& \xrightarrow[\leftarrow]{}
\underset{\underset{(A)}{H\ \ X}}{\overset{H\ \ H}{\cdot C:C\cdot}}
\end{array}
$$

$$
\begin{array}{c}
\uparrow \ \downarrow \\
H\ \ H \\
:C:C: \\
- \quad :C:C:\ + \\
H\ \ X \quad (B)
\end{array}
$$

(2-32)

In configuration (A) one pair of electrons in each monomer unit is unpaired (in the π-orbital), which enables a single electron to react with an external single electron and end up as a free radical. This is the key to the most conventional mode of polymerization, via free radicals. Configuration (B) leads to an excess of electrons on one side (anion) and a shortage on the other side (cation). This leads to ionic polymerization (cationic or anionic). Hence there are choices of various mechanisms for polymerization, where the chemical nature of the monomer (characteristics of the substituent groups) dictates the preferred mechanism. This is shown in Table 2-3.

TABLE 2-3
Initiation in Chain Polymerization

Monomer	TYPE OF INITIATION			
	Radical	*Cationic*	*Ionic*	*Stereospecific*[*]
Ethylene	+ +	+		+ +
Styrene	+ +	+	+	+
1–3 Dienes	+ +	+	+	+
Halogenic olefins, TFET	+	–	–	
Vinyl esters (VAC)	+	–	–	
Isobutylene; vinyl ether	–	+	–	
Aldehydes, ketones	–	+	+	
Butene-1, propylene	–	–	–	+
Acrylics, acrylonitrile	+	–	+	

*will be discussed later

We begin with radical initiation because it is most commonly used. In order to activate the monomers, materials that release free radicals are called on, known as the *free radical initiators*. (It should be stressed at the outset that the term *initiator* differs from the term *catalyst*, because the former becomes part of the generated molecule, as will be described later. A catalyst does not participate directly in the chemical reaction, but speeds it up.) In order to understand the mechanism of polymerization via free radical initiation, we write the relevant chemical equations.

Initiation

$$I \xrightarrow{k_d} 2R^* \tag{2-33a}$$

$$R^* + M \xrightarrow{k_i} RM^*. \tag{2-33b}$$

where
I $\ \ =$ Initiator (concentration)
$R^* \ =$ Free radical (concentration)
M $\ \ =$ Monomer (concentration)
$RM^* =$ growing chain radical

A common initiator is benzoyl peroxide BPO $(C_6H_5COO)_2$, which decomposes at 80–95°C to form free radicals, as follows:

$$C_6H_5-\overset{\overset{\displaystyle O}{\|}}{C}-OO-\overset{\overset{\displaystyle O}{\|}}{C}-C_6H_5 \rightarrow 2C_6H_5-\overset{\overset{\displaystyle O}{\|}}{C}-O^* \tag{2-34}$$

Another useful initiator is azobiisobutyronitrile (AZBN) which decomposes at 50–70°C, as follows:

$$(CH_3)_2\underset{\underset{\displaystyle CN}{|}}{C}N{=}N\underset{\underset{\displaystyle CN}{|}}{C}(CH_3)_2 \rightarrow 2\,(CH_3)_2-\underset{\underset{\displaystyle CN}{|}}{C^*} + N_2 \tag{2-35}$$

The reaction between the free radical and the monomers completes the stage of initiation:

$$(CH_3)_2\underset{\underset{\displaystyle CN}{|}}{C^*} + H_2C{=}CHX \rightarrow (CH_3)_2-\underset{\underset{\displaystyle CN}{|}}{C}-CH_2-\overset{\overset{\displaystyle H}{|}}{\underset{\underset{\displaystyle X}{|}}{C^*}} \tag{2-36}$$

It is apparent that a fraction of the initiator is retained in the chain radical, which will eventually continue to grow according to the steps of propagation and termination.

Propagation

$$. RM^* + M \xrightarrow{k_p} RMM^* \tag{2-37}$$

$$RMM^* + (n - 1)M \xrightarrow{k_p} R(M)_n M^* \tag{2-38}$$

This equation represents a whole series of reactions, wherein the monomer reaches (by diffusion) the chain radical and is attached to the growing chain. All this may happen during fractions of a second and lead to a giant chain, with a free radical at the end. A process is now needed that will terminate the large chain—the termination reaction.

Termination

$$RM_n M^* + RM_m M^* \xrightarrow{k_{t,c}} RM_{n+m+2} R \tag{2-39}$$
$$\text{(combination)}$$

$$RM_n M^* + RM_m M^* \xrightarrow{k_{t,d}} RM_{n+1}H + RM_m CH{=}CHX \tag{2-40}$$
$$\text{(disproportionation)}$$

Combination, shown in equation 2-39, occurs more frequently and leads to longer chains. In essence, during combination two growing chain radicals meet and combine to form a single chain. When the process is disproportionation, the reaction between two growing radicals yields two polymeric chains (one unsaturated).

It is obvious that all three reactions (initiation, propagation and termination) are vital for the polymerization process, but the propagation step determines both the rate of the polymerization process and the chain length (as defined later). The rate of propagation (actually representing the rate of polymerization), r_p, depends on the rate constant k_p and on the concentration of reactants. It is, in principle, identical to the rate of disappearance of the monomer.

$$r_p = -dM/dt = k_p(M)(M^*) \tag{2-41}$$

This is virtually true, since for each chain radical of size n (mers), only one monomer reacts at the initiation step, and (n − 1) monomers react during the propagation steps.

It is normally assumed (and verified experimentally) that the free radicals reach a constant concentration (steady-state assumption) so that the rate of initiation, r_i, equals the rate of termination, r_t:

$$r_I = 2 f k_d I \quad \text{(rate of initiation)} \tag{2-42}$$

where f expresses the efficiency with which free radicals formed by Equation (2-33a) react by Equation (2-33b).

$$r_t = 2 k_t (M^*)^2 \quad \text{(rate of termination)} \tag{2-43}$$

Taking $r_i = r_t$, M^* can be expressed as:

$$M^* = \left[\frac{f k_d I}{k_t} \right]^{0.5} \tag{2-44}$$

This leads to a final expression for the rate of polymerization (identical to rate of propagation).

$$r_p = k_p \sqrt{\frac{k_d f I}{k_t}} M = K I^{0.5} M \tag{2-45}$$

The rate of polymerization is proportional to the concentration of the monomer and to the square root of the concentration of the initiator.

The kinetic constants are mainly sensitive to the temperature, mostly through the Arrhenius relationship:

$$k = k_0 \exp(-E/RT) \tag{2-46}$$

where
E = energy of activation,
R = gas constant
T = temperature in Kelvin.

Equation (2-46) defines a straight-line with a negative slope, on a semi-logarithmic scale, (ln)k versus 1/T. The slope (−E/R) indicates the sensitivity to the effects of temperature. The combination of the constants in Equation (2-45) leads to a large positive value for E, which means a very high sensitivity to an increase of temperature. The rate thus increases roughly two fold to three fold with a 10°C enhancement in temperature.

How is the chain length estimated? If we define the instantaneous chain length as the ratio between the rate of propagation and rate of termination, we reach an expression for the kinetic degree of polymerization, DP, (equivalent to the number of mers in the chain) as follows:

$$DP = \frac{k_p M}{\sqrt{4fk_d k_t I}} = \frac{K'M}{\sqrt{I}} \tag{2-47}$$

Equation (2-47) holds for the case of termination by disproportionation, while when combination prevails the value should be doubled. The instantaneous molecular weight (MW) of the polymer chain is derived by multiplying DP by the molecular weight of the repeat unit.

The major difference between the expressions for rate of reaction and the degree of polymerization is that the latter depends on the inverse of the square root of initiator concentration. The reason for that is obvious, as more initiations of chains lead to shorter ones. On the other hand both rate and MW are proportional to monomer concentration. Therefore, in a batchwise process the highest rates and molecular weights occur at the outset, while later the concentration of the monomer will gradually diminish. The concentration of the initiator varies little throughout the polymerization process.

The combined constant K' differs from that for the rate function (K), resulting in a negative activation energy, so that an increase in temperature will decrease MW. This fact should be considered when controlling the temperature in the reactor, as most reactions are exothermic, and efficient ways for the absorption of the heat of reaction are essential. There are alternatives for free radical formation that come from other sources of energy such as light, heat, or high-energy radiation.

In addition to the three principal reactions that prevail in radical polymerization, a fourth one often occurs, that is, the chain transfer reaction. In this case the free radical passes from the growing chain to another molecule (usually a small one like an active solvent, the monomer itself, or even to a polymeric "dead" chain) leading to the termination of one chain together with the starting of a new one that is capable of growing.

Transfer Reactions

$$\overset{\overset{\textstyle k_{tr}}{\downarrow}}{RM_n M^* + XA \rightarrow R M_{n+1} X + A^*} \tag{2-48}$$

$XA =$ a small molecule (solvent or even contaminant)

$$A^* + M \rightarrow AM^* \quad \text{(new growth)} \tag{2-49}$$

$$RM_n M^* + M \rightarrow RM_{n+1} + M^* \text{ (transfer to monomer)} \tag{2-50}$$

An interesting result is observed. On one hand, a chain is terminated and thus shortened, while on the other, the total rate of the process may be barely affected. However, if the new radical A^* is not able to polymerize, it can only react with another radical into a simple molecule. This process called inhibition, often serves as a means of eliminating premature polymerization of the monomer during storage and transportation.

$$A^* + A^* \rightarrow A\text{-}A \tag{2-51}$$

When chain transfer occurs to a finished polymer, branches are formed. This phenomenon is found in low-density polyethylene, which exhibits various kinds of branches extending from the main chain.

$$RM_nM^* + P \rightarrow RM_{n+1}H + P^* \tag{2-52}$$

where P = polymer chain

$$P^* + M \rightarrow CH_2\overset{\overset{\displaystyle M^*}{\displaystyle |}}{-}CH_2-CH-CH_2 \tag{2-53}$$
(polyethylene with branched radicals)

In conclusion, the transfer reaction leads either to the shortening of the chain or to the formation of side branches. In both cases many physical properties will be affected, as shown later. However, it is possible to harness this reaction under controlled conditions as a chain-transfer regulator, a retarder or inhibitor. The theory of polymerization kinetics has reached a high level of development, which enables an improved engineering design of the process and good control on the rate, molecular weights and their distribution, and the existence of secondary reactions.

Ionic Polymerization   Differing slightly from the free radical mechanism that has been widely discussed, ionic polymerization uses specific catalysts to serve as electron withdrawing agents (Lewis acids and others), forming a positive ion (cation) properly named carbonium.

$$\begin{matrix} H_3C & & H_3C \\ & \diagdown & & \diagdown \\ & C{=}CH_2 & \leftrightarrow & C^+{-}CH_2^- \\ & \diagup & & \diagup \\ H_3C & & H_3C \end{matrix} \tag{2-54}$$

$$\begin{matrix} H_3C & \\ & \diagdown \\ & C^+{=}CH_2^- + H^+ \rightarrow (CH_3)_3C^+ \\ & \diagup \\ H_3C & \end{matrix} \tag{2-55}$$
(carbonium)

Whenever the catalyst serves as an electron donor (like amines), a negative ion (anion) is formed—carbanion.

$$NH_2 + M \rightarrow NH_2M^- \tag{2-56}$$

The propagation stage resembles in principle that of free radical propagation, but termination is totally different because the two chain ions bearing the same sign cause repulsion, as opposed to the attraction that is essential for termination. Therefore the mechanism of termination is much more complicated, and the growing ion has to react with a counterion or in a transfer reaction. In some cases, there are no conditions for transfer, so that the final ionic chains are still reactive (so-called "living polymers"). They may continue to grow eventually if a new monomer is provided. In practice this unique phenomenon is exploited in anionic polymerization without termination for well-defined and ordered homopolymers or copolymers.

In principle, the appearance of electron donor groups near the double bond of the monomer leads to a cationic mechanism, while positive groups that withdraw electrons mostly lead to the anionic process. An increase of temperature usually leads to a decrease in rate of reaction or length of the chain. Polymerization always occurs in solution, wherein the solvent acts as separator of the ion-pair, and quite often as a transfer reagent. As already mentioned, most monomers undergo polymerization via free radicals, mainly when conjugated double bonds are present or substitute groups that withdraw electrons.

The coordinative (stereospecific) polymerization mechanism differs from the previous ones. It caused a real revolution in the polymer world when developed in the 1950s by the scientists Ziegler (in Germany) and Natta (in Italy)—both Nobel Prize Laureates in 1963. The principle here is the use of specific catalysts that orient the mers in the chain into a highly ordered configuration. By this mechanism, ethylene forms a linear (branchless) chain, so-called high density polyethylene (specific mass 0.95 to 0.96) which was developed by Ziegler.

Propylene (and other vinyl monomers) form highly ordered structures, termed "isotactic," wherein all the side groups (methyl) appear on the same side of the plane.

$$
\begin{array}{ccc}
CH_3 & CH_3 & CH_3 \\
| & | & | \\
-CH & -CH_2-CH-CH_2-CH-CH_2-
\end{array}
\quad \text{(isotactic polypropylene)}
$$

(2-57)

Another ordered configuration is termed "syndiotactic," which is an alternating structure.

$$
\begin{array}{cccc}
CH_3 & & CH_3 & \\
| & & | & \\
-CH & -CH_2-CH-CH_2-CH-CH_2-CH-CH_2- \\
& \quad | & & | \\
& \quad CH_3 & & CH_3
\end{array}
$$

(2-58)

Both structures exhibit a high degree of crystallinity in contrast to disordered polypropylene (atactic) wherein the methyl group appears randomly in either side of the plane of the chain backbone.

$$-\overset{\overset{\displaystyle CH_3}{|}}{CH}-CH_2-\overset{\overset{\displaystyle CH_3}{|}}{CH}-CH_2-CH-CH_2-\overset{\overset{\displaystyle CH_3}{|}}{CH}-CH_2-$$
$$\underset{\underset{\displaystyle (atactic)}{\overset{\displaystyle |}{CH_3}}}{}$$

(2-59)

This latter polymer is commercially useless.

Another ordered polymer, which appears as a new homolog in polyolefin family, is the polybutylene (polybutene-1).

$$\overset{\displaystyle CH_3}{\underset{\displaystyle |}{}}$$
$$\overset{\displaystyle CH_2}{\underset{\displaystyle |}{}}$$
$$[-CH-CH_2]_n$$
(polybutane-1)

Many polymers may be formed by this novel method into highly ordered and highly crystalline chains. The catalysts developed by Ziegler and Natta are composed of a combination of organo-metallic compounds (metals from Group I–III in the Periodic Table) and halogen derivatives of transition metals (Groups IV–VIII), such as the combination $(C_2H_5)_3Al$ with $TiCl_3$. They form a complex structure that binds the monomers into a preordered position. In spite of the tremendous achievements in this field, the exact mechanism is in dispute, since the theory is complicated.

Stepwise Polymerization

As stated in Section 2.4, the condensation reaction may be the basis for the mechanism of stepwise polymerization. As long as the functionality equals 2 (di-acid and di-alcohol in polyester or di-amine in polyamide), only linear chains are obtained. However, once polyfunctionality prevails (appearance of tri-alcohol like glycerol, or tri-acid), reactive branches are formed that may interact and lead to a three-dimensional structure, called cross-linked (gelation). This is the basis for thermosetting polymers on one hand, or for stabilizing the elastomeric chain on the other hand (replacing vulcanization).

Let's designate the number of functional groups of the same type by N. If N_0 functional groups existed at the beginning of the polymerization process and, during the reaction, molecules from group A are being depleted by the counter group B, so that a single group of either type is retained at the ends of the growing chain, then:

$$\underset{(A)}{HOOC-R_1-COOH} + \underset{(B)}{HO-R_2-OH} \rightarrow$$
$$HOOC-R_1-COO-R_2-OH + H_2O$$

(2-60)

The degree of conversion (P) is expressed by

$$P = (N_o - N)/N_o \tag{2-61}$$

On the other hand, the extent of the chain (the number of repeating units = mers) is expressed as $\overline{DP}$ = average degree of polymerization.

$$\overline{DP} = N_o/N \tag{2-62}$$

The explanation is that we started with N_0 groups and ended up with N groups, all at the ends of the chains. The relationship between degree of polymerization and conversion, is derived:

$$\overline{DP} = 1/(1 - P) \tag{2-63}$$

The conclusion is that a very high degree of conversion is needed in order to gain significant molecular weights.

$$MW = DP \times M_o \tag{2-64}$$

where M_o = mass of repeat unit.

Once two monomers (with opposing functionality) interact, the average value for M_o equals half of the mass of the repeat unit. Equation (2-63) points out how difficult it will be to reach high molecular weight materials via condensation reactions. Looking, for example, at the formation of polyamide Nylon 6-6.

$$NH_2(CH_2)_6\,NHH + n\,HOOC(CH_2)_4COOH \rightarrow \tag{2-65}$$
$$H[HN(CH_2)_6NHCO(C\,H_2)_4CO]_n\,OH$$

For $P = 0.99$, $DP = 100$ (from Equation 2-63).

$$M_o = 1/2\,HN(CH_2)_6HNCO(CH_2)_4CO = 113, \text{ therefore, } MW = 11,300.$$

In reality, the highest obtainable MW is around 50,000. In addition to the requirement of a very high conversion (which also calls for a long time) it is crucial to avoid a surplus of any of the reactants or the presence of mono-functional groups and contaminants. All of these may nullify a functional group and just substitute it—thus making this end of the growing chain inactive. This requires strict efforts for equivalent functionality (stochiometric ratios). Any side product (mostly water) must be removed, otherwise a reverse reaction may operate (depolymerization).

Copolymerization

We have been discussing the mechanism of polymerization of a single monomer (homopolymer). However, when two or more monomers polymerize in situ, the chains obtained differ in composition and location of each mer. These are controlled by the ratios between the tendency to homopolymerize and to cross-polymerize r_1 and r_2. (Details will follow.)

A random chain constructed by monomers A and B via an addition reaction looks as follows:

AAABBAAABAAAABBBABAAABB

(It is important to stress that the polymers obtained via condensation are not copolymers, as their repeat unit is identical.)

The importance of the reactivity parameters r_1 and r_2 is revealed if the four basic reactions for a copolymer chain (by free radicals, as an example) are expressed as follows:

$$(1)\ A^* + A \xrightarrow{k_{11}} AA^*$$

$$(2)\ A^* + B \xrightarrow{k_{12}} AB^* \qquad (2\text{-}66)$$

$$(3)\ B^* + B \xrightarrow{k_{22}} BB^*$$

$$(4)\ B^* + A \xrightarrow{k_{21}} BA^*$$

Reactions (1) and (3) are identical to homopolymerization. The parameters r_1 and r_2 are defined as the ratio between the kinetic constant for homopolymerization and that for cross-reaction.

$$r_1 = k_{11}/k_{12} \qquad r_2 = k_{22}/k_{21}$$

The rate of disappearance of monomer A $(-dA/dt)$ will be derived from Equation (2-66) (1) and (4), while that of B $(-dB/dt)$ is derived from reactions (2) and (3). By dividing both rates, the ratio dA/dB is obtained, and that describes the instantaneous composition ratio of the two mers in the chain. Let's call this ratio $dA/dB = Y$, whereas the appropriate ratio of molar concentration in the mixture of the monomers will be termed, $X = A/B$. The expression for the composition in the chain Y (ratio) as related to the ratio of the monomer feed (X) represents the so-called "equilibrium equation."

The composition of the feed varies in a batch reaction, but keeps constant in a continuous one. In any case, the instantaneous composition of the copolymer (Y) and monomer feed (X) is expressed as:

$$Y = \frac{X(r_1 X + 1)}{r_2 + X} \tag{2-67}$$

Equation (2-67) has a deficiency in graphical presentation, as the ratios Y and X are bounded between zero and infinity. Therefore, it is better to normalize by using mole fractions in the polymer:

$$dA/(dA + dB) = F_1 = Y/(1 + Y)$$

and in the monomer mixture

$$f_1 = A/(A + B) = X/(1 + X)$$

In a binary system (copolymer)

$$f_1 + f_2 = 1$$

$$F_1 + F_2 = 1$$

The boundaries for mole fractions are 0 and 1. Equation 2-67 is transformed to:

$$F_1 = \frac{f_1(r_1 f_1 + f_2)}{f_1(r_1 f_1 + 2f_2) + r_2 f_2^2} \tag{2-68}$$

While both Equations 2-67 and 2-68 are very useful, the graphical presentation of Equation 2-68 is advantageous, (as shown in Figure 2-1).

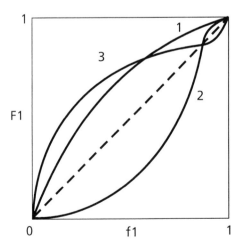

FIGURE 2-1 Equilibrium curve for copolymerization

The diagonal $(F_1 = f_1)$ serves as the reference line. This line represents the unique case where $r_1 = r_2 = 1$, which leads to $Y = X$ in Equation 2-67, or $F_1 = f_1$ in Equation 2-68. In this unrealistic case, the composition of the copolymer is identical to that of the monomer mixture.

Another extreme case is described wherein $r_1 = r_2 = 0$, which means that none of the monomers is polymerizable by itself, and yet a copolymer is obtained. Equations (2-67) and (2-68) verify in this case that $Y = 1$ and $F_1 = 0.5$ (always 50% of each component). This structure, named alternating, may be identified upon examining the chain:

ABABABABAB

as there is no way to produce AA or BB bonds. The most common structure is the random chain

AAABBAABAAABBA

where the exact composition will be dictated by r_1 and r_2.

The intersection between the equilibrium curve and the diagonal is called the azeotropic point (resembling that in binary distillation) in which $Y = X$ or $F_1 = f_1$. The conditions for an azeotrope may be found by using Equation (2-67).

$$Y = X = \frac{1 - r_2}{1 - r_1} \tag{2-69}$$

which requires that both r_1 and r_2 are either smaller or larger than 1 (see curves 2 and 3 in Figure 2.1). Curve 1 applies when $r_1 > 1$ and no azeotrope exists. In this case, the copolymer is always richer in A as compared to the composition of the monomer feed mixture. If r_1 and r_2 are quite large, one obtains block copolymers,

AAAAAABBBBBBAAAAA

Another configuration is called a *graft copolymer*, wherein one monomer appears as a branch on the main chain, consisting of the other

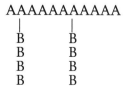

Needless to say, by copolymerization a wide variety of structures and compositions may be derived. Copolymers are widely used in elastomers and

artificial textiles, as in many plastomers. The most common are SAN (styrene-acrylonitrile); SB (styrene-butadiene); EP (ethylene-propylene); Saran® (vinylchloride-vinylidene chloride); or ABS (a terpolymer comprised of acrylonitrile, butadiene and styrene).

2.5 POLYMERIZATION—INDUSTRIAL PROCESSES

Transfer of the polymerization process from the laboratory to a full scale industrial level calls for the solution of many engineering problems—the structure of the reactor, removal of heat and thermal control, mixing of a high viscous mass, etc. In spite of the advantages of a continuous production line, there are still batch reactors in use, mainly for small scale production. In the condensation polymerization process, some aspects must be carefully followed—exact stochiometric balance between reactants (and prevention of loss of either monomer); elimination of by-products (water or other small molecules) sometimes by employing vacuum; prevention of decomposition by oxidation at high temperatures (use of an inert atmosphere); and controlling premature gelation.

These processes are frequently batchwise, while viscosity increases gradually up to the final stages. In most cases separation between monomers and polymer is unnecessary. On the other hand, polymerization processes based on free radical initiation involve strongly exothermic reactions that call for critical control of temperature and removing heat from highly viscous systems. In most cases, separation of excess monomer is essential as well as polymer separation from the dispersing fluid.

If possible from the point of view of efficiency, polymerization of the reactants without any use of a dispersing medium is desired. This method is called Bulk Polymerization. However, in order to absorb the heat released in the reaction, other methods like suspension or emulsion polymerization are frequently used. In this case, the water absorbs heat and provides good control of temperature. Another (less popular) method is the solution (homogeneous or heterogeneous) polymerization, which is mostly utilized in ionic initiation. We compare advantages and disadvantages of each method. (Table 2-4 describes the most useful polymerization methods for some commercial polymers).

Bulk Polymerization

This method is simple to operate, efficient in regard to reactor space utilization, and leads to a polymer that is pure without the retention of other species. On the other hand, the absorption of heat is quite difficult, leading to the danger of overheating that prevents controlled production. There is

TABLE 2-4
Methods of Industrial Polymerization

Polymer	Bulk	Solution	Suspension	Emulsion
PVC	+	+	+ +	+ +
PS	+ +	+	+	+
PE	+ +	+	+	+
PP		+ +		
Acrylics	+ +	+	+	+
ABS				+ +
Nylon	+ +			

also a high energy demand for mixing a stirred reactor, as a result of the very high viscosity. In a tubular reactor, comprised of many small diameter tubes, the absorption of heat is easier, but polymerization temperature is less constant, causing a wider molecular weight distribution. Overheating may also lead to the phenomenon of self-acceleration. This method is mainly used in condensation processes, but is also adopted in the production of polyethylene and PVC, as well as in monomer casting operations in molds with acrylics.

Solution Polymerization

This method enables good control of temperature while the viscosity of the system drops. It is also sometimes advantageous to utilize the polymer product in solution, as with coatings and paints. On the other hand, both chain length and rate of production decrease, as a result of the diminished concentration of the monomer. Other problems are related to solvent retention and recovery, and hazards of fire and health.

Suspension Polymerization

This is a popular method (mainly in PVC) that improves control, while circumventing some of the disadvantages of the former methods. The product is obtained as "pearls" that are handy for further processes. In suspension polymerization some additives (stabilizers) are needed, enabling the dispersion of monomer droplets and preventing coagulation. When stirring stops, the polymer precipitates. The reaction occurs in the monomer droplets in which the initiator is dissolved. Efficient drying and elimination of dispersing additives are crucial.

Emulsion Polymerization

This is an important method that, in addition to improved thermal control, results in a higher rate of production, higher molecular weights and a narrow distribution. The product may be utilized as an emulsion like rubber latex or coatings. Otherwise the emulsion must be destroyed, and the polymer carefully dried and separated, because retention of residues from the emulsifier (detergent) may cause the deterioration of properties (mainly for electrical uses).

The kinetics of emulsion polymerization differ completely from that of bulk or suspension. The soap (detergent) is used not only as a stabilizer but mainly as locus of the polymerization—so-called soap micelles. These consist of an array of 20–100 molecules of soap creating a micellar structure of 25–50 Å in length (radius). The initiator is water soluble and the free radicals and the monomer molecules diffuse into the hydrophobic interior of the micelles while water is attracted to the hydrophilic exterior zone. Thus, the micelles serve as the core of growing polymer particles. At a later stage the micellar structure disappears, and the process continues in the polymer particles swollen by monomers, leaving the soap as a protective layer on the particle surface. The concentration of soap dictates the molecular weight and rate of production.

PROBLEMS

1. Define the term "chain length." If two polymers have the same chain length, do they also have the same molecular weight?

2. What is the molecular weight of PVC which has 750 mers?

3. Is polyester a copolymer?

4. In the case of a polymer made by condensation, is the repeating unit identical to the sum of the monomers?

5. What is the molecular weight of a cross-linked polymer. How are cross-linked polymers dissolved?

6. In a polymerization process where there is no change in the viscosity

during initial 2 hours and then a rapid rise occurs, what is the mechanism? What is happening in the first two hours?

7. In a polymerization process where the viscosity rises slowly for several hours and then stabilizes, what is the mechanism? What is the significance of measuring the viscosity?

8. A petrochemical plant manufactures polyethylene by free radical initiation in a continuous process. Solve the following changes:
   a. Increase the rate of polymerization by 50%, without changing the monomer concentration. How does it affect the molecular weight?
   b. Increase rate of polymerization by 100% and molecular weight by 20%. Hint: change concentrations of both monomer and initiator.
   c. Decrease molecular weight without changing concentrations. Offer two options with no calculation.

9. What is the significance of termination in addition polymerization? What happens when a transfer reaction replaces the termination? What happens when neither termination nor transfer occur?

10. A polymer is manufactured from the monomer $HO(CH_2)_4COOH$.
   a. What is the generic name of the polymer?
   b. What is the mechanism of polymerization?
   c. What is the molecular weight when conversion is 95%, 98%, 99%?
   d. Is it possible to reach 100% of conversion?

11. A copolymer is made from vinyl chloride and vinyl acetate, so that the molar concentration of the acetate in the polymer is 5%.
What is the composition of the feed?
Plot an equilibrium curve.

      Data: vinyl chloride (1)   vinyl acetate (2)
                   $r_1 = 1.4$ $r_2 = 0.65$

12. Plot an equilibrium curve and modify the basic equation for the following cases of copolymerizations:
   a. $r_1 = r_2 = 1$
   b. $r_1 = r_2 = 0$
   c. $r_1 = r_2$

13. Compare the following polymerization processes and define the difference in the product properties. The polymer is PVC via free radical initiation.

    a. Bulk polymerization.
    b. Suspension polymerization.
    c. Emulsion polymerization.
    d. Solution polymerization.

3

Structure and Characterization of Polymers

3.1 THE CHEMICAL STRUCTURE

THE CHEMICAL STRUCTURE of the homopolymers consists of an exact repetition of the chemical structure of the mer unit. The chemical bonds are mostly primary covalent ones, mainly $C-C$ and $C-H$, but include also $C-O$, $C-N$, and so forth. There appear also various isomers, double bonds (unsaturation), tertiary or quaternary carbons, ring-like structures, and others. There are also secondary bonds, albeit weaker than the primary ones. One typical isomer, found in polymers, is based on the presence of "head-to-tail" bonds as compared to "head-to-head" or "tail-to-tail" bonds. This is illustrated by the vinyl group, with the substitute radical $-X$.

$$CH_2-CH-CH_2-CH-CH_2-CH- \quad \text{(head-to-tail structure)}$$
$$\quad\quad\ \ | \quad\quad\quad\ \ | \quad\quad\quad\ \ |$$
$$\quad\quad\ X \quad\quad\quad X \quad\quad\quad X$$

or

$$-CH_2-CH-CH-CH_2-CH_2-CH-CH-CH_2-$$
$$\quad\quad\ \ | \quad\ | \quad\quad\quad\quad\quad\ | \quad\ |$$
$$\quad\quad\ X \quad X \quad\quad\quad\quad\ X \quad X$$

(tail-to-tail and head-to-head structure)

The head-to-tail isomer occurs more frequently. Other types of isomers may be defined under the heading of chain architecture. The nature of the substituent group dictates the polymer behavior, mainly through secondary intermolecular bonds or rotation around this bond. That determines the flexibility of the chain, its mechanical strength, and the chemical affinity (compatibility) to other compounds (solvents, plasticizers and other additives). Taking for example the substituent group of the halogens (X = F, Cl, Br, I), it is well known that the strength of the primary bond diminishes from fluorine to iodine, respectively. Hence, the fluorocarbon polymers present excellent thermal and chemical stability (including weather resistance). Polyvinylchloride is affected mainly by its polar secondary bond and is, therefore, more rigid than polyethylene, while the polymers based on bromine or iodine are unstable. The bromine derivatives, however, are most efficient as fire retardants because of their weaker bonds. Strong secondary bonds result from the existence of polarity (dipole-moment, as in the case of PVC) or from hydrogen bridges (as in the case of polyamides).

$$-HN-(CH_2)_6-N-\overset{\overset{\displaystyle O}{\|}}{C}-$$

(hydrogen bridge)   H   Nylon 6-6

$$-HN-(CH_2)_6-NH-\overset{\overset{\displaystyle O}{\|}}{C}(CH_2)_4-\overset{\overset{\displaystyle O}{\|}}{C}$$

The basic chemical structure, including end groups, obviously determines the chemical behavior, that is, resistance to various chemicals such as acids, bases, salts, and organic solvents. As an example, polyamides (Nylon) are sensitive to acidic hydrolysis because of their chemical structure. On the other hand, the existence of hydrogen bridges produces a compact, highly crystalline structure, with high mechanical strength. A stiff structure is observed when aromatic or heterocyclic rings are incorporated in the chain skeleton, and when weak C—H bonds are replaced by the stronger C—O and C—N bonds. This is the background for the synthesis of novel polymeric families that withstand high temperatures (so-called HT polymers).

3.2   THE MOLECULAR ARCHITECTURE

The molecular architecture defines the geometry of the chain. In general, the polymer chain is described by a coil that is capable of adopting various forms in accordance with the ambient conditions, solvents, and temperature. Under conditions where there is no net interaction with the surroundings (so-called theta state, or temperature), a model of a randomly condensed coil

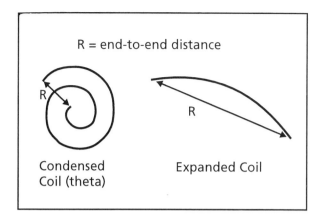

FIGURE 3-1 Polymer chain dimensions

is postulated which is capable of opening and expanding with temperature rise or the use of a better solvent. Hence, the same chain may comprise various volumes under different conditions, where the linear dimension is best represented by the end-to-end distance. This must be taken into account in the case of molecular weight determination when measuring the viscosity of dilute solutions. In rare cases, the chain may be represented by a stiff rod, mostly found in bio-polymers (like DNA and others). See Figure 3-1.

According to the steric structure there can be three major configurations—a linear chain, a branched chain, or a cross-linked (network or 3-dimensional) chain, as shown in Figure 3-2. The linear chain has no branches or chemical links with neighboring chains (albeit with the existence of physical links, called entanglements).

The branched chain contains side subchains, mainly obtained by the chain transfer mechanism. Most branches are short but yet are able to decrease the crystallinity in the polymer, due to steric hindrance. Existence of longer branches affects the flow properties by decreasing viscosity. This

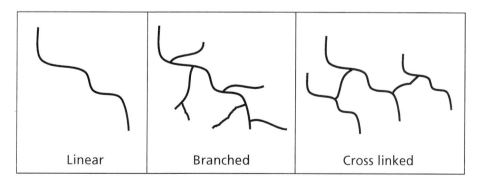

FIGURE 3-2 Steric structure of chains

FIGURE 3-3 A zigzag structure of PE

improves workability both in melt or solution. However, extremely long branches (of the same order of magnitude as the main backbone) can reverse the effect by increasing viscosity caused by entanglements between adjacent branches. A similar effect is observed when the branches are reactive, such as polyfunctional groups in a condensation reaction which lead to primary intermolecular ties. This represents the third steric structure, the cross-linked chain. Here a lattice network structure is obtained by primary bonds. By increasing the degree of cross-linking, the chain gradually loses its mobility and capability to dissolve and it finally reaches the thermosetting state. In this case, the molecular weight and viscosity reach "infinity." In intermediate conditions there appear soluble and nonsoluble phases, which may be separated, named sol and gel, respectively. In true cross-linking, there appears to be no dissolution, but the gel may swell under the effect of solvents. Needless to say, only linear or branched chains show flow properties (thermoplastics).

The highly ordered planar structures, as obtained by stereospecific polymerization, were mentioned in Section 2.3. In contrast to the branched chains, we deal here with linear chains where side substitute groups appear regularly on the same side of the plane, or alternatively. These structures are named isotactic or syndiotactic, respectively, and the degree of order is termed "tacticity." In both cases, good conditions for crystallization exist, even when the polymer itself is known as difficult to crystallize (as in the case of polystyrene or polymethylmethacrylate). Any orderless structure is named atactic.

Another look at the geometric structure of polymer chains may show ordered domains like helical coils, wherein the bond and rotation angles are predetermined. Isotactic polypropylene, as an example, possesses a rotation angle 120°, so that the structure repeats itself after each three mers. In simpler cases, a zigzag structure prevails with a constant bond angle (mostly tetrahedrons with an angle of 109.5°), as in polyethylene (See Figure 3-3).

A rotational angle between 0° and the 180° prevails in most cases, thus affecting the practical dimensions of the chains, namely, the end-to-end distance. The most remote distance appears in the trans configuration, and shortest distance is in the cis configuration. The stiffness of the chain and the ambient conditions (temperature) affect the configuration of the chain and, as a result, its properties.

3.3 MOLECULAR WEIGHTS AND THEIR DISTRIBUTION

High molecular weight represents the most important parameter of macromolecules. One can never expect, however, to find chains of uniform size in industrial manufacture. In reality, chains of various lengths are present, so that only an average molecular weight can be determined. Full characterization would require the width of the molecular weight distribution, as well.

This makes the distinct difference between the meaning of molecular weight in the case of macromolecules and the constant unique values of molecular weight for small molecules. Another complexity comes from the choice of the mode of averaging. As is well known in statistics, there are many ways of averaging, while in a heterogeneous ensemble each average has a different value.

Taking, for example, a "simple" polymeric chain such as polyethylene, the number n of repeating units (mers), defines the "chain length." In polymer science, this is termed Degree of Polymerization (DP).

$$(CH_2-CH_2)_n \qquad DP = n$$

In order to express molecular weight (MW) of the polymer, the DP is multiplied by the molecular weight of the repeating unit. In this case it is $(CH_2-CH_2) = 28$. So, if we have DP = 1,000, then MW = 28,000. That would be the case, if we could assume a single unique value for DP. As mentioned before, however, that does not happen in most practical cases.

From all the averaging methods, it is customary to use the number-average, $\overline{M}_n$, and weight average, $\overline{M}_w$, which are both well defined in a mathematical way.

$$\overline{M}_n = \frac{n_1M_1 + n_2M_2 + \ldots + n_NM_N}{n_1 + n_2 + \ldots + n_N} = \frac{\Sigma n_iM_i}{\Sigma n_i} = \frac{\Sigma W_i}{\Sigma \dfrac{W_i}{M_i}} \qquad (3\text{-}1)$$

where n_i = number of molecules (or moles), of constant MW = M_i. This method actually counts the number of chain groups of definite length.

$$\overline{M}_w = \frac{W_1M_1 + W_2M_2 + \ldots + W_NM_N}{W_1 + W_2 + \ldots + W_N} = \frac{\Sigma W_iM_i}{\Sigma W_i} = \frac{\Sigma n_iM_i^2}{\Sigma n_iM_i} \qquad (3\text{-}2)$$

where W_i = weight of a chain that has MW = M_i. The relationship between weight (W_i) and number (n_i) of the chains is simply expressed by

$$W_i = n_iM_i.$$

It is easy to prove that, by definition, $\overline{M}_w \geq \overline{M}_n$.

Therefore, the ratio $\overline{M}_w/\overline{M}_n$ may serve as a measure of the breadth of the distribution. It is customary to prefer the use of mole fractions (x) or weight fractions (w), that sum to unity $\Sigma x_i = \Sigma w_i = 1$. A general expression may represent various averages, as follows:

$$\overline{M} = \frac{\Sigma n_i M_i^x}{\Sigma n_i M_i^{x-1}} \tag{3-3}$$

This enables the use of some higher moment averages like $\overline{M}_z$ and $\overline{M}_{z+1}$.

$$x = 1; \overline{M} = \overline{M}_n \quad \text{(number average)}$$
$$x = 2; \overline{M} = \overline{M}_w \quad \text{(weight average)}$$
$$x = 3; \overline{M} = \overline{M}_z \quad \text{(z average)}$$
$$x = 4; \overline{M} = \overline{M}_{z+1} \quad \text{(z + 1 average)}$$

While the two higher moment averages (z, z + 1) are more sensitive to the existence of "very long tails" only the first two averages ordinarily have physical significance. In order to stress the difference between the two major averages, let us look at a numerical (but artificial) example. Suppose that a polymer consists of 10^3 chains of $M_1 = 10^6$, and 10^3 chains of $M_2 = 10^4$. Then

$$\overline{M}_n = \frac{10^3 \cdot 10^6 + 10^3 \cdot 10^4}{2000} = 505000$$
$$(W_1 = 10^9; W_2 = 10^7) \quad W_i = n_i M_i$$
$$\overline{M}_w = \frac{10^9 \cdot 10^6 + 10^7 \cdot 10^4}{10^9 + 10^7} = 990200 \cong 990000$$

In this example, the difference between the two averages stems from the fact that, in the case of $\overline{M}_w$, the chains are weighed, while for $\overline{M}_n$ they are only counted. It is obvious that $\overline{M}_w$ must be sensitive to the very long chains (that weigh more) and $\overline{M}_w$ is closer to M_1 (the higher MW group). The ratio $\overline{M}_w/\overline{M}_n$ = 1.96, which is close to the "statistical" value of 2.0. This molecular weight distribution (MWD) is considered to be narrow, whereas in practice a high value (broad distribution) is sometimes met. (For example, up to 40–50 for low density polyethylene, LDPE.)

Taking a binary mixture of the same weight, say 1kg of each group, the new $\overline{M}_n$ will be different.

$$\overline{M}_n = \frac{\Sigma W_i}{\Sigma \dfrac{W_i}{M_i}} = \frac{2}{\dfrac{1}{10^6} + \dfrac{1}{10^4}} = 19800. \quad \text{whereas} \quad \overline{M}_w = 505000$$

It is clearly understood that $\overline{M}_n$ is more sensitive to the existence of the short chains (the less heavy).

We can now plot a "normal" distribution curve, as W(M), which means weight fraction per unit molecular weight, versus M. This provides us with the best information regarding the shape of MWD and the exact location of each average. In this curve, we incorporated another average (actually a range of averages), named $\overline{M}_v$ (viscosity average molecular weight). See Figure 3-4.

$\overline{M}_v$ is determined by measuring the viscosity of a polymer dilute solution, and using appropriate correlations between the so-called intrinsic viscosity of the polymer in the solution $[\mu]$ and the molecular weight. This is defined as follows:

$$[\mu] = \lim_{c \to 0} \frac{\mu - \mu_s}{\mu_s C} \tag{3-4}$$

μ = viscosity of solution at concentration C.
μ_s = viscosity of the pure solvent.

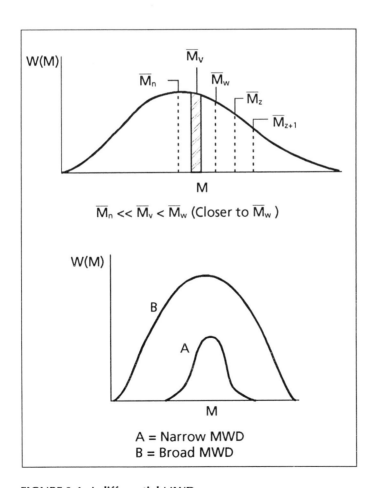

FIGURE 3-4 A differential MWD curve

The value of intrinsic viscosity is reached upon extrapolation to infinite dilution (C = 0).

The empirical relation between $[\mu]$ and M is named after Mark and Houwink (the M-H equation):

$$[\mu] = K \cdot \overline{M}_v^a \tag{3-5}$$

Such an expression may be linearized on a log-log scale, where a = the slope of the straight line. The common range for the exponent, a, lies between 0.5 and 0.8.

Finally, the viscosity average molecular weight, $\overline{M}_v$, is defined:

$$\overline{M}_v = \left(\frac{\sum n_i M_i^{a+1}}{\sum n_i M_i} \right)^{1/a} \tag{3-6}$$

When a = 1, $\overline{M}_v = \overline{M}_w$, so that $\overline{M}_v$ is usually close to $\overline{M}_w$, but its value depends on the parameter a, and hence one observes a range of $\overline{M}_v$ (as shown in the graph, Figure 3.4). However, $\overline{M}_v$ will approach $\overline{M}_n$, only if one predicts a hypothetical value for a = -1. The parameters K and a, for each pair of polymer-solvent (at a given temperature), are found by calibrating with narrowly distributed polymer samples (fractions), for which M (prefer $\overline{M}_w$) is given. Much data can be found in handbooks.

The intrinsic viscosity (IV) is physically related to the "hydrodynamic volume" of the chain (its volume during flow), so that when the chain dimensions (as defined by end-to-end distance) increase in the solution, the IV will increase en suite. In practice, a large deviation is found between IV in a "bad" solvent (worst conditions at the so-called theta (Θ) state, where a = 0.5), and in an "excellent" one (where high values of IV are found and a = 0.8).

The theta state is defined for the polymer-solvent system (only in case of poor solvents) as that temperature where minimal interaction between the chain and its environment prevails. This is also defined as the threshold limit of dissolution (at high molecular weight) and the chain takes on a configuration of a dense coil.

The correlation between IV and the chain dimension in solution is best expressed by Flory's universal expression:

$$[\mu] = AR^3/M = BV/M \tag{3-7}$$

In this equation A and B represent universal constants, R represents a typical linear dimension of the chain (usually the end-to-end distance), and V = hydrodynamic volume of the chain in solution. One may assume that the product $[\mu]$ M is associated with the volume of the chain. This assumption will be undertaken in a later discussion of MWD utilizing a size separation technique (GPC).

The measurement of intrinsic viscosity is performed in a capillary viscometer (mostly made of glass) which determines the flow times of the pure solvent and those of the polymer solution at various concentrations (in the diluted region). The ratio of flow times, between the polymer solution and the pure solvent, approximates the viscosity ratio (relative viscosity, μ_r)

$$\frac{t}{t_s} = \frac{\mu}{\mu_s} = \mu_r \qquad (3\text{-}8)$$

$$\frac{\mu - \mu_s}{\mu_s} = \mu_{sp} = \mu_r - 1 \qquad (3\text{-}9)$$

The term μ_{sp} (specific viscosity) expresses the relative contribution of the polymer to the viscosity of the solution.

It is customary to measure the viscosity of a polymer melt (at high temperature), μ_o, which is very sensitive to MW.

$$\mu_o = A\overline{M}_w^{3.4} \qquad (3\text{-}10)$$

Here, μ_0 represents melt viscosity under low shear conditions (the Newtonian range).

More details about the viscosity are described in Chapter 4, which deals with rheology of polymers. The industry prefers to measure melt fluidity (inverse viscosity), by using a standard rheometer, named Melt Flow Indexer. The measured property (mass of polymer that flows through a standard capillary under determined load and temperature, for 10 minutes) is also known as MFI (Melt Flow Index). Because MFI is inversely related to melt viscosity, it is important to remember that low values of MFI are obtained by high MW polymers.

There are various experimental methods for the assessment of MW averages. We will describe some that are absolute and need no calibration. The number average, $\overline{M}_n$, can be obtained, in principle, by utilizing common methods of general and organic chemistry—elevation of boiling point, suppression of melting point, and reduction of vapor pressure (the so-called colligative properties). The major deficiency stems from the fact that when increasing the molecular weight it becomes more difficult to measure the diminishing changes in temperature or pressure. If, for example, a low concentration of 0.01 molar is used, one needs a solution of 100 g/L for M = 10^4 and 1,000 g/L for M = 10^5. At these concentrations it is impossible to obtain a polymer solution instead of a gel. On the other hand, further reduction of the molar concentration reduces the measured reading.

Sometimes, it is possible to use the simple method of "neutralization" of a reactive chemical end group (like COOH in polyester or polyamide). Upon increasing molecular weight, however, the concentration of the reactive end groups diminishes significantly. Therefore, the use of these methods is limited to the range of $\overline{M}_n$ < 20,000. Above that value, it is customary to mea-

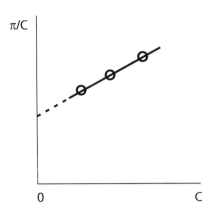

FIGURE 3-5 Osmotic pressure of polymer solutions

sure the osmotic pressure formed when a dilute polymer solution is separated from the pure solvent by a semipermeable membrane. The simple relationship between the osmotic pressure (π) and the other variables (molecular weight M, concentration C, and temperature T) is described by the next expression. See Figure 3-5.

$$\frac{\pi}{C} = RT \cdot \left(\frac{1}{M} + BC \right) \tag{3-11}$$

The coefficient B (so-called second virial) depends on the thermodynamic interaction between the polymer and the solvent. It vanishes at the theta condition, which leads to the Van't Hoff equation, commonly used in general-chemistry $\pi = (CRT)/M$.

The same expression is obtained after extrapolation to C $=$ 0, the limiting absolute value for $\overline{M}_n$. The number average molecular weight, as measured by this method, is frequently utilized in polymer science, in the region of $\overline{M}_n$ $= 2 \cdot 10^4 - 5 \cdot 10^5$. Below this region, a false reading may result, due to the possible passage of short chains through the membrane. In order to avoid this discrepancy, good selectivity of the membrane is essential.

The weight average is obtained by measuring the light scattering in polymer dilute solution at various concentrations in a photoelectric cell. High clarity and purity of the solution is crucial for obtaining accurate results. The observed parameter is the "turbidity" coefficient, τ, which is related to MW according to the following expression:

$$\frac{HC}{\tau} = \frac{1}{M} + 2BC \tag{3-12}$$

Constant B is identical to that in Equation (3-11). H represents a combination of optical parameters, fixed for the system. Again, the constant B vanishes at theta condition, eliminating the need of measuring at several concentrations. As mentioned, this method is very sensitive to contaminants and dust, and it is often necessary to measure at several angles and at various concentrations.

Except for research laboratories, it is uncommon in industry to utilize the exact absolute methods for molecular weight estimation. Instead, it is common to measure the viscosity of a single concentration solution, or the melt, usually the MFI. In quality control, one is concerned with fluctuations rather than absolute values.

Since the mid 1960s, a new instrument has been developed called GPC (Gel Permeation Chromatography), which provides the molecular weight distribution curve as well as all the averages. This method is based on liquid chromatography, giving exact information about the concentration of chains of varying length, according to the elution time, after flowing in dilute solution through packed columns filled with well-defined porous gels. See Figure 3-6.

It is interesting to note, that the shortest chains (including the solvent itself) undergo the longest route through the internal pores while the longest chains (unable to penetrate into the gel pores) are rejected and emitted earlier. By appropriate calibration, the chromatogram is converted to a true (finger-print) molecular weight distribution curve, from which all pertinent averages are easily calculated. The most reliable calibration is based on the relationship between the product $[\mu]M$ (intrinsic viscosity multiplied by molecular weight) and elution volume (or time), following Flory's equation (3-7) which relates intrinsic viscosity to hydrodynamic volume and molecular weight. This may be described by the "universal calibration curve," a straight line on a semilogarithmic scale. See Figure 3-7.

In research and development the characterization of polymers, as regards to molecular weight averages and distribution, is essential in order to predict

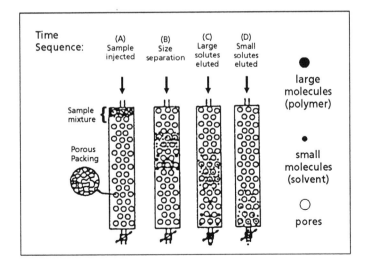

FIGURE 3-6 Separation in GPC columns

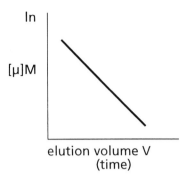

FIGURE 3-7 Universal
calibration curve for GPC
measurements

the relationship between the molecular structure and polymer performance. In industry it is important to follow polymer characterization in view of controlling the homogeneity in production and deriving improved performances. In general, many properties are improved by an increase in molecular weight but it becomes difficult to process polymers exhibiting very long chains. A typical example is the UHMW (ultra high molecular weight) polyethylene used for very thin plastic bags. Enhanced mechanical properties can be obtained, but at the price of convenience in fabrication. As for the width of distribution, it is well known that widely distributed polymers gain better processability, albeit with some drop of mechanical properties.

Most physical properties (including viscosity) are related to the weight average, but some are correlated to the number or intermediate averages. It is advantageous to obtain the complete molecular weight distribution data from which the various averages and general equations may be derived (all facilitated by a good computer).

3.4 TRANSITIONS

The performance of polymers may be best understood by considering the concept of "thermal transition" that determines the transition between phases and states. The primary transition, between a crystalline solid phase and an amorphous liquid (polymers never appear as gases), is defined by the melting point, T_m. In this case, genuine changes in the primary thermodynamic properties (enthalpy, specific volume) occur, as with small molecules. The major difference is that in polymers there is no sharp transition point but a range of temperatures — depending on molecular weight distribution and on the degree of crystallinity (discussed in Section 3.5). In essence, T_m defines the disappearance of the crystalline phase, being the temperature at which the last crystallites melt. It increases with an increase of molecular weight, concentration and dimensions of the crystallites. It is affected by the rigidity of chemical structure of the chain core, presence of strong intermolecular bonds, and bulky side groups. Needless to say, amorphous polymers lack the primary transition and, therefore, have no melting point. They soften upon heating, however, and solidify glasslike at low temperatures. The amorphous phase, however, exhibits a secondary transition temperature of extreme significance, the so-called glass transition temperature, T_g. It is named secondary (in spite of primary importance), due to the fact that a step-change of secondary thermodynamic properties occurs (such as thermal expansion coefficient or specific heat), representing the temperature derivatives of primary functions. Therefore, T_g corresponds to a change in the slope of the thermodynamic function with temperature, as shown by the plot in Figure 3-8.

A glass transition occurs for an amorphous polymer when the slope of a plot of specific volume changes abruptly with rise of temperature. The slope itself (the thermal expansion coefficient α), $\alpha = (1/V)\, dV/dT$, shows a step-change at T_g. A crystalline polymer also shows a transition at the same temperature, due to its amorphous phase. (It will be explained later that only semi-crystallinity may exist and there are almost always amorphous domains.) In the case of a crystalline polymer, however, a primary transition temperature, T_m, exists, signified by an abrupt change in volume. The specific volume of the crystalline polymer is always lower than that of the equivalent amorphous one because its more compact structure increases the density.

Many physical properties change in the vicinity of T_g (it is really a region), the most prominent being the dynamic ones. A transition between two dynamic states occurs in the amorphous phase—between the "glassy" state, wherein the mobility of chain fragments is frozen (frozen "free volume") and the elastic state wherein the chain mobility increases upon the rise of temperature (the free volume increases and viscosity decreases, respectively). (Free volume is defined as the difference between the volume of the liquid phase and the extrapolated value at absolute zero temperature.) It is customary to define the transition (T_g) when a fraction of frozen free volume of 2.5% appears, and stays constant at lower temperatures. There is no "real" solidification, however, but a frozen liquid. The rate of the measurement (or the fre-

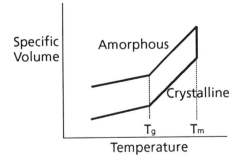

FIGURE 3-8 Transition temperatures

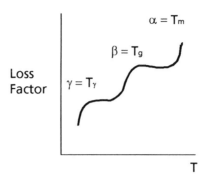

FIGURE 3-9 Multiple transitions

quency) is of importance, showing that the transition is basically dynamic in nature, a state of relaxation. It is therefore advised to gain accuracy by measuring at a low rate of temperature changes, as the values of T_g increase with higher rates of measurements. Actually, some more transition temperatures may be found (both below and above T_g) by applying a dynamic test under a wide range of temperatures. If one plots the "loss factor" (described in Section 4.1) in a dynamic test (mechanical or electrical), several peaks appear, describing very clearly the existence of transitions. The upper one (only in the case of a semicrystalline polymer) represents T_m, and then T_g, but also T_γ, which is related to the release of some degrees of freedom, or mobility, at a lower temperature. See Figure 3-9.

The existence of T_γ (or other transitions) is associated with the rotation of side groups or other well defined fragments of the mer, thus releasing some additional frozen free volume at temperatures below T_g. It is obvious, that the various transition temperatures have a paramount effect on the polymer performance. Around T_g many parameters are drastically changing, such as

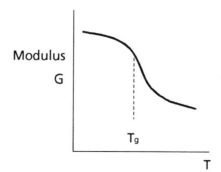

FIGURE 3-10 Changes in modulus around T_g

the modulus of rigidity, G, (see Figure 3.10), coefficient of diffusion, viscosity, and light refraction index. It is believed that the melt viscosity of polymers at T_g is about 10^{13} poise (very high).

Above T_m amorphous liquid appears, while a semi-crystalline solid appears below T_m. Above T_g, the amorphous phase shows a high mobility leading to high elasticity and deformity. On the other hand, below T_g most polymers behave like rigid solids, and are frequently brittle (glass-like). Extensivity is drastically reduced, together with the capacity to resist the effect of blows and oscillation (impact strength).

Therefore, polymers that have a low T_g (compared to service temperature) behave rubber-like or leather-like, with high toughness. Most prominent are the polyolefins and rubber (both synthetic and natural). Other polymers exhibit relatively high values for T_g, like PS, PVC, PMMA, PC and others. In principle, these polymers are predicted to perform at room temperature as rigid, with low deformability and with a tendency to brittleness. Some are exceptions, however, like PC (polycarbonate) which is very tough in spite of the existence of a rather high T_g (around 150°C). It has been postulated that the expected brittleness at room temperature is eliminated due to the appearance of another transition temperature, T_γ at (−100°C).

For most engineering polymers a high T_g (above 100°C) is required, whereas elastomers (synthetic rubbers) need low T_g (usually below −50°C). It is also possible to reduce T_g by external plasticification (addition of plasticizers which generate a solid solution with organic liquids of very low transition temperatures) or by polymerization with other comonomers which leads to lower transitions.

Empirical mixing equations may be useful in this case, like:

$$\frac{1}{T_g} = \frac{W_1}{T_{g1}} + \frac{W_2}{T_{g2}} \tag{3-13}$$

(Note: T_g must be expressed at absolute temperature.) W_1 and W_2 represent weight fractions of the appropriate components in the mixture or in the copolymer, respectively. By computing the additivity of free volumes it is possible to gain higher accuracy. It is important to note that T_g, in a way similar to T_m, is affected by various parameters—molecular weight, degree of crystallinity, rigidity of the chemical structure, and secondary bonds. It is therefore not surprising to find a narrow range for the ratio (in absolute temperatures) T_g/T_m between 1/2 and 2/3 for many polymers. This enables initial estimation of a transition temperature if only one of the two is given. There are many exceptions, however, to this simple empirical rule. It is also possible to predict T_g of a new polymer, from its chemical structure, by adding contributions of various structural units. It is interesting to observe the dependence of T_g on the chemical structure, as shown by the following cases:

Polyethylene, PE
$-CH_2-CH_2-$
$T_g = -120°C$

Polypropylene, PP
$-CH-CH_2-$
$\underset{|}{CH_3}$
$T_g = -17°C$

Polyvinylchloride, PVC
$-CH-CH_2-$
$\underset{|}{Cl}$
$T_g = +87°C$

Polystyrene, PS
$-CH-CH_2-$
$\underset{|}{C_6H_5}$
$T_g = +100°C$

Therefore, PS is a rigid (high modulus) but a brittle polymer, because of the presence of a bulky phenyl side group. It can be modified by an elastomeric phase (usually by copolymerizing), thus augmenting its impact strength. PVC is also rigid with a low impact strength, unless it is modified. It is customary, however, to plasticize PVC by external plasticizers or by copolymerization (with vinyl acetate, for example). The large difference between the values of T_g for PE and PVC, is related to the existence of polar forces in PVC.

In conclusion, the existence of various transition temperatures indicates different states in the matter, where T_m and T_g represent the more important ones having thermodynamic significance. T_g is still considered the most dominant one, dictating the mechanical and dynamic behavior. Its values appear in various equations as "corresponding states," such as universal parameters at identical T/T_g or $(T - T_g)$. This idea will be developed further in a discussion on viscoelasticity. (See Section 4.1). Table 3-1 exhibits transition temperatures of some selected polymers.

TABLE 3-1
Transition Temperatures of Some Polymers

Polymer	$T_g°C$	$T_m°C$
Polyethylene	-120	137
Polypropylene	-17	176
Polybutene-1	-25	125
Polyisobutylene	-73	
PVC	87	212*
Polystyrene	100	240*
Polycarbonate	150	250
Nylon 6-6	50	265

*Only for isotactic (crystalline).

3.5 MORPHOLOGY

Morphology deals with the structure in the solid state. There is a distinct indication that in principle regular polymers are apt to crystallize—meaning to form long-range spatial order. There is an apparent difference, however, between the crystallization of long chains and that occurring in low molecular weight organic or inorganic compounds. In polymers, a complete crystalline structure is never reached, as noncrystalline (amorphous) regions always interfere. Therefore it is common to use the term semi-crystallinity and to characterize it by "degree of crystallinity," defined as the fraction (percentage) of the crystalline phase in the total mass. The theory of polymer structure is a prominent branch of polymer physics, which has mainly developed since the 1950s. It should be mentioned that around 1920 the concept of long chains had been introduced, when X-ray diffraction studies provided the proof for the existence of an ordered crystalline structure. (The intensity of the obtained diffraction rings is measured and the distances apart are transformed into a defined lattice conformation.) It was tacitly assumed that the long chains create a macroscopic structure wherein some regions exhibit order while others do not. (See Figure 3.11a.) The same long chains pass through the crystalline phases (A) and the amorphous ones (B), alternately. This was described by the so-called "fringed micelle," dominant for a long period. Nowadays, it is considered obsolete—single crystals have been isolated that exhibit an almost perfect structure. The pioneering discovery in this field appeared in 1957, but arguments among various researchers as to the representative morphology still exist today.

The new concept that has been accepted refers to a structure of "folded chains." In electron microscopy, a structure of lamelli is obtained that looks like pyramids, wherein, the longitudinal axes of the chain appear perpendicular to the surface of the lamella (of about 100 angstrom width). On the other hand, calculation of the length of chains of common polymers leads to values of the order of 10^4 angstrom and up, requiring a model of folded chains. See Figure 3.11b. Even with this model, 100% crystallinity cannot be reached, as the surfaces of folds, ends of chains, and other structural dislocations contribute to the existence of an amorphous phase. The polymers that evidence the morphology of single crystals are those that can reach a high degree of crystallinity, (such as PE, PP, acetals and others), working with dilute solutions under well controlled conditions.

There are many deviations from this perfect model, such as multiple passes (like switchboard), varying dimensions of chain folds (see Figure 3.11c) and even the micelle structure in the case of low crystallinity. Unfortunately, there is no direct visible validation of the real morphology.

Crystallizing from the melt leads to a combined morphology consisting of amorphous domains (tie lines) connecting zones of single crystals. See Figure 3-12.

A polymer to be crystallizable must have a regular self repeating structure, with spatial order.

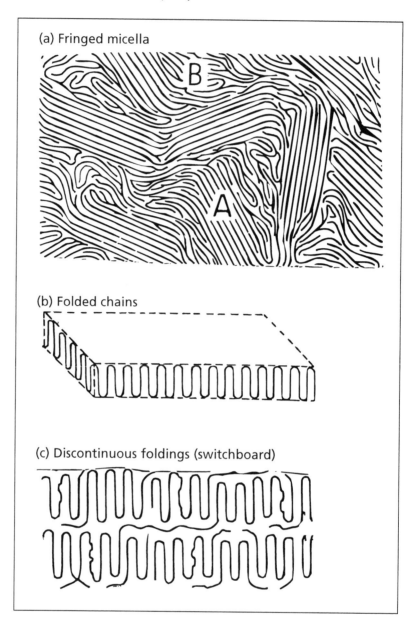

FIGURE 3-11 Morphology of polymer chains

In order to speed-up the rate of crystallization a chemically simple and symmetric structure is preferred, exhibiting high mobility and lacking steric hindrances (branching, cross-linking, bulky side groups, and structure disorder). A random copolymerization always diminishes the chance for crystallinity, up to nil. Accordingly, the following polymers crystallize easily—polyethylene (the linear reaches a higher degree than the branched), isotactic polypropylene, polybutylene, polytetrafluoroethane (Teflon), polyacetal, saturated polyesters, and polyamides.

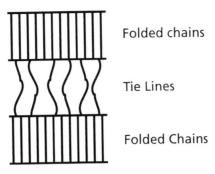

Folded chains

Tie Lines

Folded Chains

**FIGURE 3-12 Morphology of
crystallites from polymer melts**

Typical representatives of amorphous polymers are polystyrene and PMMA, which lack crystallinity due to steric hindrance by side groups. They can appear as semi-crystalline, however, after stereospecific polymerization into tactic configurations. On the other hand, polycarbonate does crystallize but the process takes a long time, calling for carefully controlled annealing conditions. A modern approach suggests that some structural order persists in the amorphous state, albeit to a much lesser degree than in the crystalline state.

Let us describe the kinetics of this process. As in any crystallization, the first step consists of nucleation, where nuclei are formed by cooling the melt (or solution) below the melting (solidification) temperature. This is a reversible process when nuclei are formed and destroyed. When a critical size is obtained, these nuclei remain as the centers for the growth of crystallites. Nucleation is most rapid, therefore, as the temperature drops; the determining factor being the degree of supercooling $\Delta T = T_m - T_c$ (between the melting point T_m and the temperature of crystallization T_c).

The growth of the crystallite represents the second stage of the process which may be followed with a polarizing optical miscroscope. Under ideal conditions, macroscopic spherical structures (spherulites) are apparent. They grow in all directions, according to external conditions. The radius of the spherulite may represent the length dimension, where high supercooling creates many small spherulites. Crystallization at higher temperatures (approaching the melting point) produces fewer spherulites (fewer nuclei) but of larger dimensions.

The rate of crystallization differs from one polymer to another, mainly depending on temperature and chain length. Under the most suitable conditions prevailing for each individual polymer the degree of crystallization is dependent on the chain structure and the annealing period. The effect of the temperature on the rate of crystallization is verified by a maximum between the two major transition temperatures, T_m and T_g. See Figure 3.13.

The rate can be increased by applying other external modifications, solvents, plasticizers, or nucleation promoters. The extension of chain length

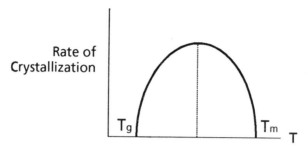

FIGURE 3-13 Kinetics of crystallization

up to some limit favors crystallization, but upon increasing chain length (or molecular weight), the mobility of the chain diminishes and eventually decreases the rate of growth. This leads to a maximum. It is possible to assist the crystallization process by applying external stresses, such as by stretching rubbers or by inserting frozen orientation in other polymers. In the latter case, a crystalline and selectively oriented region can be obtained which is highly favored in man-made fibers. Here, a stretching stage provides high mechanical strength in the direction of orientation.

Determination of the degree of crystallinity can be achieved by using several methods. Though not very accurate, the simplest way concerns the specific volume, V, (or its inverse value, the specific mass). The degree of crystallinity is calculated, using data for the estimated specific volumes of the pure amorphous phase V_a and of the crystalline lattice cell V_c.

$$f = \frac{V_a - V}{V_a - V_c} \qquad (3\text{-}14)$$

where f = degree of crystallinity, $V = 1/d$, and d = observed density. An improved precision is obtained from measurements of X-ray diffraction or calorimetry (DSC) that provide information on transition temperatures and enthalpy of melting, or by NMR.

The rate of crystallization (the time dependence) was calculated by Avrami for the initial stage (up to 30% of crystallinity, approximately) following the expression:

$$f = k \cdot t^n \qquad (3\text{-}15)$$

The power n is postulated to have integral values between 1 and 4. The value of (n) correlates to the some of the number of crystallite dimensions (3 for a spherulite, 2 for a disc, and 1 for a needle shape) and the order of the nucleation step (0 for spontaneous heterogeneous process, and 1 for homogeneous nucleation rate which is proportional to time, t).

As an example, a homogeneous nucleation step involving sphere-like crystallites (spherulites) would predict n = 4 (maximum). In reality, many exceptions, and nonintegral values may be found. At higher degrees of crystallinity, impingement between growing crystallites cannot be disregarded. This leads to the general equation:

$$1 - f = e^{(-kt^n)} \qquad (3\text{-}16)$$

where the constants k and n are identical to those of the simplified Avrami equation.

It is important to note that the existence of crystallinity has a crucial effect on many chemical and physical properties. In general, upon increasing the crystallinity, resistance to the attack of chemicals and solvents, as well as to hostile environments, is augmented. In practice, it is difficult to dissolve a crystalline polymer in appropriate solvents unless temperatures approaching the melting point are used. From the mechanical point of view, the rise of crystallinity increases rigidity and strength (mostly determined by the increase of modulus and yield strength) but at the same time decreases elongation and impact strength. The optical properties are also affected by the presence of crystallites, as the transparency decreases. Small spherulites are more favorable than larger ones for improving optical and mechanical properties.

In closing, crystallinity is most essential for engineering structures. An additional stage of orientation is usually implied for fibers, by stretching and freezing the ordered structure.

PROBLEMS

1. A mixture is made of two mono-dispersed polystyrenes at 50% (weight). The molecular weight of the first sample is 2.5×10^6 and that of the second is 4×10^5.
   a. Calculate the weight-average molecular weight of the mixture.
   b. Calculate the number-average molecular weight of the mixture.
   c. Determine the distribution.
   d. Calculate the viscometric-average molecular weight from a solution in a theta solvent; in toluene (a = 0.75)

2. The viscosity of a solution of polyisobutylene (PIB) was measured at 30°C. For a polymer concentration of 0.2 gram per 100 cc the viscosity was

found to be 1.6 centipoise. The viscosity of the pure solvent (cyclohexane) at the same conditions was 0.8 cps. Calculate the following:

  a. The intrinsic viscosity of PIB in cyclohexane at 30°C.
  b. The viscometric average molecular weight.
  c. Assuming that $\overline{M}_v$ is very close to $\overline{M}_w$, and that the distribution is normal, calculate (π/c) as c → 0.

MH equation: $[\eta] = K\overline{M}_v^a$

Huggins equation: $\dfrac{\eta_{sp}}{C} = [\eta] + k[\eta]^2C$

Data:
A. The constant k in Huggins equation is 0.4.;
B. The constants for the MH equation are $K = 2.6 \times 10^{-4}$ a = 0.70. Concentration is always expressed in g/dl.

3. The curve for MWD of a PMMA may be expressed as:

$W(x) = ax^2 + bx$

where $x = DP$; $W(x) =$ weight fraction at $DP = x$; and $b = 6 \times 10^{-6}$.
  a. Draw the full distribution curve, and determine the values of x at the ends and at the peak.
  b. Determine $\overline{M}_n$, $\overline{M}_v$ (at Θ), $\overline{M}_w$, $\overline{M}_z$
  c. Express the distribution ratio.
     (Hint: Use integration instead of summation.)

4. Suggest methods for determining molecular weight $(\overline{M}_n)$, for the following polymers:
  a. Nylon 6-6, estimated $\overline{M}_n = 4,000$
  b. PMMA, estimated $\overline{M}_n = 100,000$
  c. PMMA, estimated $(\overline{M}_n = 5,000$

5. The following data were obtained during a fractionation analysis of a commercial polymer. Calculate molecular weight averages and degree of dispersion.

Fraction number	Weight in grams	Molecular weight
1	1.5	2,000
2	5.5	50,000
3	22.0	100,000
4	12.0	200,000
5	4.5	500,000
6	1.5	1,000,000

6. A GPC apparatus is calibrated by PS standards (narrow distributed) dissolved in TCB at 130°C.
   a. Draw the universal calibration line and write its equation.
   b. What will be the molecular weight of a fraction of linear polyethylene (HDPE) that has a peak at V_e = 30.

Data:
A. GPC data for calibration
   elution volume V_e = 32.5   M = 50,000
   elution volume V_e = 34.8   M = 10,000
B. MH constants for PS in TCB at 130°C
   K = 8.95×10^{-5};   a = 0.727 (C in g/dl)
   for HDPE at same T and solvent,
   K = 5.96×10^{-4};   a = 0.70

7. The viscosity of a dilute solution of PMMA in chloroform is measured in a capillary viscometer at 25°C. Determine the intrinsic viscosity.

Concentration g/dl	Flow time seconds
0	120
0.20	155
0.40	196
0.60	243

8. For the polymer sample described in problem 7, calculate the viscosity-average molecular weight. Data for PMMA samples which are narrowly distributed show the following intrinsic viscosities in chloroform at 25°C.

MW	Intrinsic viscosity
10^4	0.054
10^5	0.34
10^6	2.145

9. Two samples of polyethylene show the following characteristics:
   A. Specific weight 0.920 MFI = 20
   B. Specific weight 0.930 MFI = 2.0
      a. Which sample has a higher molecular weight? Estimate the ratio.
      b. Which sample has a higher intrinsic viscosity?
      c. Which sample will appear first in a GPC?
      d. Which sample is more crystalline?

10. Nylon has T_g = 50°C. Estimate its melting point.

11. Toothpaste is shipped to the following countries: Zambia (lowest temperature in winter +5°C), and Iceland (lowest temperature in winter −60°C). The paste is packed in plasticized PVC tubes.
      What will be the plasticizer concentration in each case?
      Data: T_g of pure PVC = +87°C; T_g of PVC containing 10% (by weight) of plasticizer is +40°C.

12. ABS is a terpolymer made of acrylonitrile (A), butadiene (B) and styrene (S). The glass transition temperatures (T_g) are as follows: for A, T_g = +104°C; for B, T_g = −90°C; for S, T_g = +100°C. Two copolymers are produced: copolymer (1) consisting of 70% (weight) of styrene and the rest acrylonitrile; copolymer (2) consists of butadiene 80% (weight) and the rest is acrylonitrile. The terpolymer is obtained by mixing the two copolymers.
      a. Estimate the final composition of the terpolymer that will have T_g = −10°C.
      b. Is it possible to reach the same T_g (−10°C) with a copolymer of styrene and butadiene? If yes, at what composition?
      c. Are either of copolymers (1) or (2) considered to be an artificial elastomer? Comment?

Note: For copolymers and mixtures use the mixing rule, W_i = weight fraction:

$$\frac{1}{T_g} = \frac{W_1}{T_{g1}} + \frac{W_2}{T_{g2}}$$

13. During crystallization of a polymer, it was found that the nucleation is spontaneous and the crystallites are needlelike.

    a. Using Avrami's law, estimate the degree of crystallinity after 10 minutes. (It has been observed that 10% crystallinity is reached after 4 minutes.)

    b. Calculate the density of the polymer, if the density of the crystallite phase is 1.0 and that of the pure amorphous phase is 0.86.

    c. Repeat the problem, if spherulites appear after 10 minutes.

14. Half life time for crystallization of acetal has been found to be 27.7 minutes. At 4 minutes after crystallization started, 10% of crystallinity was obtained. What is the shape of the crystallite? What is the type of nucleation?

4

Behavior of Polymers

*I*N THIS CHAPTER we will discuss extensively the behavior of polymers and their response to stress. This will allow us to better understand the physical and thermal properties—relating structure and performance.

4.1 RHEOLOGY OF POLYMERS

Rheology is a relatively young branch of natural science that deals with the relationships between forces (stresses) and deformations of material bodies. Hence, it is also connected to the flow properties of polymers both in solution and in the melt, as well as the reaction of materials in the solid state to mechanical stresses. Most polymeric materials exhibit the combined reactions of both liquid and solid states, called viscoelasticity, a combination of the viscosity of a liquid and the elasticity of a solid.

First of all, let's start with the problem of polymer flow. This also happens to be an important practical issue, because, during shaping and processing, polymers must undergo fluid flow (the "plastic" state). Long chains (which are not cross-linked) may slide and flow like any other liquid—when in solution or in a state that allows motion (above T_g, in the case of an amorphous polymer, or above T_m for a crystalline polymer). Compared to simple liquids, polymers are very different and have extremely high viscosity and a special flow characteristic, which is termed "non-Newtonian." It is therefore appropriate to state Newton's law, in which the coefficient of viscosity, η, appears. The flow process is defined as an irreversible deformation of the body, caused by shearing forces that change the shape and location of material particles. Newton's law in flow is expressed as:

$$F/A = \tau = \eta\dot{\gamma} \tag{4-1}$$

where τ represents the shear stress (shear force per unit area) that acts on the fluid, and $\dot{\gamma}$ represents the rate of shear (or velocity gradient) that expresses the rate of deformation (or strain rate).

This is illustrated in Figure 4-1.

$$\text{Note: } \gamma = \frac{\ell}{L}; \quad U = \frac{d\ell}{dt}; \quad \frac{d\gamma}{dt} = \dot{\gamma} = \frac{d\ell}{Ldt} = \frac{U}{L}$$

The physical interpretation of the unit of viscosity (η) is defined as equivalent to the force needed to affect the flow of a fluid that is bounded between two solid parallel plates (one stationary and the other moving with the fluid) while variables (distance between the plates, velocity and cross-section area) are unity.

Only in the case of simple liquids, a constant viscosity coefficient (η) may occur, dependent only on temperature. This dependence is mostly described by a typical Arrhenius equation:

$$\eta = A \exp\left(\frac{E}{RT}\right) \tag{4-2}$$

It is obvious that the viscosity of liquids drops exponentially with the temperature elevation. E expresses an activation energy for flow, and its size dictates the sensitivity to temperature changes.

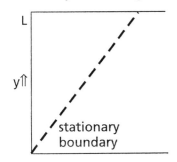 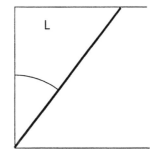

FIGURE 4-1 Shear stresses and deformation

In the case of polymers, it has been found that the viscosity is not a constant, but varies with the flow conditions. Therefore, these liquids are termed non-Newtonian. In Figure 4-2 we show the relationship between shear stress and shear rate (the so-called "flow curve"). The Newtonian liquid is represented by a straight line from the origin (1) while there may be several curves (2, 3, 4) representing non-Newtonian liquids.

In the straight line, 1, describing the Newtonian liquid, the viscosity is identical to the slope. In curve 2, the viscosity gradually diminishes with increasing shear rate. This represents a shear-thinning (pseudoplastic) liquid, typical of most polymeric melts and solutions. Curve 3 describes a shear-thickening (dilatant) case, wherein the viscosity increases with an increase of shear rate. This may appear in concentrated pastes. Curve 4 is actually described by a straight line on a linear plot, but it does not start at the origin. This is a "Bingham liquid," wherein a threshold value of shear stress, τ_o, appears, below which no flow occurs (this behavior is shown by toothpastes).

It is obviously impossible to provide a single mathematical expression for all prevailing non-Newtonian liquids. It is very popular in engineering, however, to use a power-law equation, describing a straight line on logarithmic scales;

$$\tau = K\dot{\gamma}^n \tag{4-3}$$

where
$n < 1$ represents a shear-thinning liquid,
$n = 1$ represents the Newtonian case, and
$n > 1$ represents the shear-thickening liquid.

Needless to say, the Bingham liquid is also non-Newtonian. This may be shown by calculating the "apparent viscosity" μ, defined as the ratio between shear stress and rate of shear.

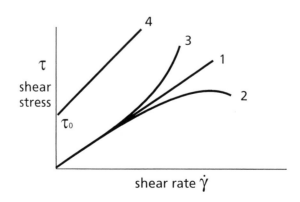

FIGURE 4-2 Flow curves of various liquids

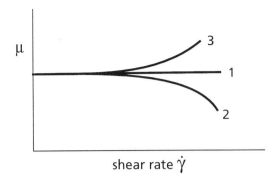

FIGURE 4-3 Apparent viscosity of various liquids

$$\mu = \frac{\tau}{\dot{\gamma}} \qquad\qquad (4\text{-}4)$$

Only in the case of a Newtonian liquid does $\mu = \eta$ (that is, the viscosity is a constant independent of shear rate, or stress). It is, therefore, advisable to plot apparent viscosities versus shear rate as in Figure 4-3.

Here, again, (as in Figure 4-2) system 1 represents a Newtonian liquid (constant viscosity); system 2 is the shear-thinning case (viscosity drops); while system 3 represents the shear-thickening case (viscosity increases). Only when conditions of low shear prevail may one consider the use of the initial constant viscosity η_o, which can be correlated with molecular weight (MW). See Figure 4-4. A typical empirical power law is followed by many polymers. That is,

$$\eta_o = KM^b \qquad\qquad (4\text{-}5)$$

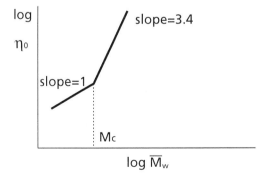

FIGURE 4-4 Dependence of melt viscosity on MW

The slope b (on a logarithmic scale) approximately equals unity at low values of molecular weight (below a critical value M_c); above M_c it is mostly around 3.4. This sharp transition of the low-shear viscosity is related to the existence of physical coupling between the growing chains, usually defined as *entanglements*. In this case, which is frequently observed, the 3.4 exponent expresses a very high sensitivity to an increase of chain length (molecular weight). In these equations, the weight-average is preferred, although the molecular-weight distribution is also controlling. The critical value, M_c, differs with the nature of the repeat unit. In reality, molecular weights are usually above the critical value, so that the following relationship is very useful:

$$\eta_o = K\overline{M}_w^{3.4} \tag{4-6}$$

(This equation can also be derived by applying modern flow theories which will not be discussed in this introductory text—reptation.)

Shear thinning itself may be attributed to a disentanglement of chains, orientation in the flow direction, and structure deformation—all resulting from increasing shear stress or rate. The viscosity, as a result, will drop. It is important to note that these factors dominate with increase in chain dimensions, so that one has to extrapolate experimental data to the region of low shear rates to gain reliable correlations. Various equations of state may describe the flow-curves, using the concept of a low-shear constant (Newtonian) viscosity parameter η_o. (B, b, n, c in next equations 4-7 through 4-9 are constants.) See Figures 4-5 and 4-6.

$$\text{Ferry: } \frac{1}{\eta} = \frac{1}{\eta_o} + B\tau \tag{4-7}$$

$$\text{Gee and Lyons: } \frac{1}{\eta} = \frac{1}{\eta_o} + b\tau^n \tag{4-8}$$

$$\text{Carreau: } \eta = \frac{\eta_o}{(1 + b\dot{\gamma})^c} \tag{4-9}$$

The Carreau equation becomes a power-law equation at high shear rates.

In conclusion, the viscosity of polymer melts depends on shear conditions (rates or stresses), on the molecular weights, and on the temperature. While Newtonian liquids obey an Arrhenius type dependence on temperature, on the other hand, polymer melts follow suit only at temperatures that exceed 100°C above the glass transition temperature (T_g). At the intermediate range, a generalized WLF equation (named after its founders Williams, Landel and Ferry) is applicable:

$$\log_{10} \frac{\eta_T^o}{\eta_{T_g}^o} = -\frac{17.4(T - T_g)}{51.6 + T - T_g} \tag{4-10}$$

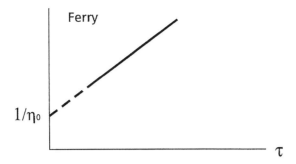

FIGURE 4-5 Ferry's equation

where,
η_T^o = low-shear viscosity at T.

The WLF equation gives the ratio of η_T^o to viscosity at T_g, that is, $\eta_{T_g}^o$ which is estimated to be 10^{13} poise (at low MW and shear). The WLF equation is considered to be an important equation of state, corresponding to the temperature difference $(T - T_g)$.

In practice, it is advantageous to be able to decrease the viscosity (thus decreasing the resistance to flow) by increasing pressure or velocity, as well as to modify the flow geometry in order to increase shear rates. In some cases, there is an effect of the history of previous deformation. This is exemplified by thixotropic liquids that show a low viscosity after a shear history (like stirring), while a high viscosity prevails at rest. This phenomenon is much desired in the case of paints and varnishes.

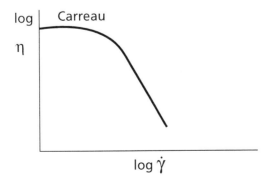

FIGURE 4-6 Carreau's equation

In non-Newtonian liquids there appears another factor, the existence of "normal stresses." In contrast to shear stresses, these represent stresses in the same direction as the deformation plane, resulting in "stretching" of the liquid and swelling of the melt extruded from a tube or die. This is actually an elastic contribution to the deformation of the body, which does not exist in simple (Newtonian) liquids. The normal stresses also cause "climbing" of the polymer melt on a stirrer when polymer solutions or melts are mixed.

In spite of the complicated behavior, the parameter of viscosity is still considered to control the flow regime, but one should be very cautious in picking the proper value, preferably at low shear. However, industry often uses pseudo-parameters that are not accurately defined, like the famous useful MFI (melt flow index), that was previously mentioned. This parameter is used frequently to characterize industrial polymers. In spite of its high sensitivity to molecular weight changes, chain branching, and distributions, it is used to represent "fluidity" (inverse viscosity) at some standard shear stress. It is, therefore, shear dependent and cannot serve as a constant parameter. It may however be modified, by extrapolating data measured at various shear stresses, in order to correlate to molecular weight. Needless to say, the MFI parameter is very useful in industry, for a simple and quick quality control means (possibly on line)—to maintain the homogeneity of production.

We recall that the combination of both liquid and solid behavior is termed *viscoelasticity*. We have already discussed the basic law for the "simple" liquid, Newton's law. For solids, Hooke's law defines the relationship between stress S and deformation γ, using a material parameter called the modulus of elasticity: $S = G\gamma$. G represents a shear modulus, while E represents Young's modulus in tension ($S = E\gamma$). That is the behavior of so-called "ideal elastic solids." This may occur mostly in metals or rigid materials, while in the case of polymers, Hooke's correlation is found only in the glassy state, below T_g.

The shear rate that appears in Newton's law is identical to the time derivative of the deformation γ: $\dot{\gamma} = d\gamma/dt$, so that, while a solid responds instantaneously to stresses (deformation independent of time or history), a liquid responds according to the rate of deformation, therefore, deformation is time-dependent. In conclusion, viscoelastic materials exhibit a dual dependence on stress or deformation as well as on time; $S = f(\gamma, t)$.

A convenient physical interpretation may be illustrated by simulating mechanical or electronic models. In the mechanical simulation, a spring represents an elastic or Hookean solid (modulus), while a piston moving in an infinite cylinder filled with a viscous liquid (a dash-pot) represents the Newtonian liquid (viscosity). Thus, the deformation of the solid (spring) is completely recoverable, while that of the liquid (dash-pot) is irrecoverable and is converted to heat. See Figures 4-7, 4-8, 4-9. In conclusion, the elastic energy is conserved and recovered while the viscous energy is dissipated.

Linear combinations of these models (linear viscoelasticity) may represent various viscoelastic states. The most common ones are the Maxwell visco-

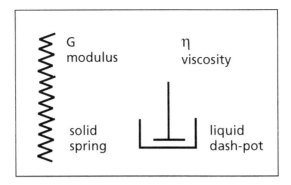

FIGURE 4-7 Mechanical models for solids (spring) and liquids (dash-pot)

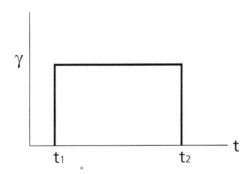

FIGURE 4-8 Deformation of an elastic solid (at constant stress)

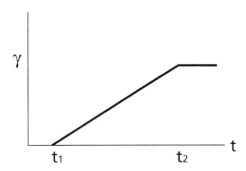

FIGURE 4-9 Deformation of a Newtonian liquid (at constant stress)

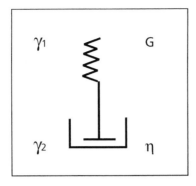

FIGURE 4-10 A Maxwell model

elastic liquid (a series combination) or the Voigt (Kelvin) viscoelastic solid (a parallel combination), as shown in Figures 4-10 through 4-13.

In Figure 4-10:

$$S = S_o;$$

$$\gamma = \gamma_1 + \gamma_2; \quad \gamma_1 = \frac{S_o}{G} \text{ (spring)};$$

$$\dot{\gamma}_2 = \frac{S_o}{\eta} \text{ (dash-pot)}.$$

The development of the deformation γ with time in a Maxwell model (when a constant stress is applied and later removed) is shown in Figure 4-11.

Here the figure shows, at time t_1 a certain stress S_o is applied on the system, causing an immediate deformation γ_1 of the solid (the spring in the model) in addition to a constant-rate deformation γ_2, representing the response of the liquid (the dash-pot in the model). γ_2 can actually continue to increase, but the forces applied to the system are suddenly removed at

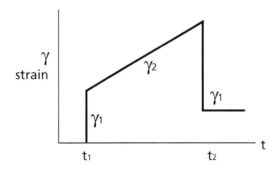

FIGURE 4-11 Deformation of the Maxwell model

time t_2. The typical response of this viscoelastic body, after removal of the stress, is as follows: the solid (spring) retracts instantaneously to its primary state, so that all the elastic deformation γ_1 is recovered. However, the liquid is unable to recover any of its deformation and as a result the contribution of the dash-pot is unchanged (an irreversible state). Altogether, the Maxwell model represents a viscoelastic liquid, where the final behavior is that of a liquid that incorporates an elastic element. It is apparent that the characteristic responses of an elastic solid and a viscous liquid are simply superimposed.

In quite another case, a body is instantaneously stretched and then kept at constant strain, γ_0. In the Maxwell model, the spring has an immediate response and tends to retract to its original position while the dash-pot will continue to move, keeping the combined constant deformation. During this period of time the stresses relax from the initial high stress developed by the spring to zero stress when the spring is no longer stretched (the liquid alone cannot support the stress). This is shown in Figure 4-12.

A typical material time-constant, namely the relaxation time λ, will govern the relaxation process.

$$S = S_o e^{-t/\lambda} \qquad (4\text{-}11)$$

where $S_o = G\gamma_0$ (response of the spring), and $\lambda = \eta/G$ (relaxation time).

λ increases with an increase in chain length. In the case of the Maxwell model, the relaxation of stress is complete, while with real materials, residual stresses that are unrelaxed may appear. It is also customary to express the stress relaxation process, by describing the time-history of the relaxation modulus G.

$$G(t) = S(t)/\gamma_0 \qquad (4\text{-}12)$$

This has a similar profile to that shown in Figure 4-12.

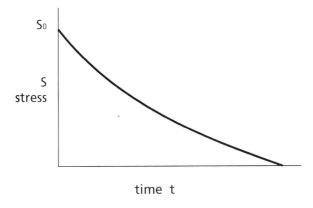

FIGURE 4-12 Stress relaxation of a Maxwell model

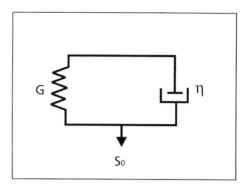

FIGURE 4-13 A Voigt (Kelvin) model

Another viscoelastic model is derived, by combining a spring and a dash-pot in parallel. This is named after Voigt (or Kelvin) and shown in Figure 4-13.

$$\gamma = \gamma_1 = \gamma_2$$
$$S_o = S_1 + S_2$$

(Numbers 1 and 2 refer to the spring and the dash-pot, respectively.)

A constant load (stress S_o) is applied and as a result the whole body will stretch. However, the spring cannot respond immediately as it is connected to the dash-pot in a rigid manner. Therefore, there will be a retarded but final elastic deformation (damped). On removal of the load, the deformation will reverse, still retarded (time-dependent). The strain γ is identical in both elements, but the stress is divided between them. This is shown in Figure 4-14.

The initial deformation is characterized by a positive slope that gradually diminishes, eventually reaching a plateau which is determined by the ultimate deformation of the spring. If at t_1 the load is removed, the deformation

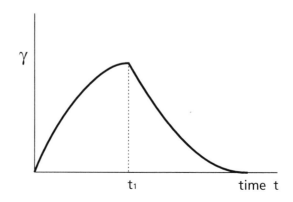

FIGURE 4-14 Deformation of a Voigt body

at this stage will recover completely but gradually. This model is therefore considered to represent the total recovery of a viscoelastic solid.

These responses are met in engineering practice as creep and recovery, and are considered to be of paramount importance when dealing with the mechanical performance of plastic materials under an extended period of static loading. As in the case of stress-relaxation, this behavior is controlled by a characteristic material parameter called the retardation time $(\lambda = \eta/G)$. The history of creep and recovery may also be traced by a time-dependent compliance, $J(t) = 1/G(t)$.

$$J(t) = \gamma(t)/S_o \qquad (4-13)$$

Real materials exhibit a much more complex behavior compared to these simplified linear viscoelastic models. One way of simulating increased complexity is by combining several models. If, for instance, one combines in series a Maxwell and a Voigt model, a new body is created, called the Burger model (Figure 4-15).

We choose to separate the spring and the dash-pot from the Maxwell element, which does not affect the result. There are four basic parameters: G_1, G_2, η_2 and η_3. On stretching, due to application of a load, the response may be attributed to three elements—the single spring, the Voigt element and the single dash-pot. The response of the single spring is instantaneous, according to Hooke's law. This represents the glassy state—a quick response in a short time or a low temperature. The Voigt element contributes retarded elasticity and the combination acts as an elastic solid susceptible to creep under load. The single dash-pot is a flow element, making the combined body a viscoelastic liquid with some portion of unrecovered deformation γ_2. All this is shown in Figure 4-16.

The Burger model represents the behavior above T_g in the elastomeric state, while the single dash-pot simulates melt flow at higher temperatures. Around the transition between the glassy and elastomeric states, the modulus of elasticity drops 10^3 to 10^4 fold, and the material becomes flexible and

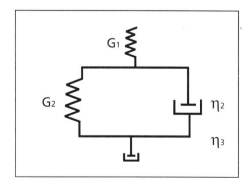

FIGURE 4-15 The Burger model

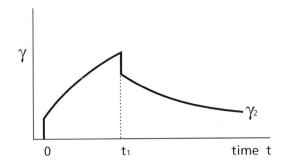

FIGURE 4-16 Deformation of a Burger body

ductile. The effects of temperature or time are similar. In a short time (high speed), only the contribution of the solid (the single spring of the model) will prevail. At larger times, the retarded elastic response is added, while at very long time (steady state) the contribution of the single dash-pot (melt flow) will dominate. By eliminating the isolated dash-pot from the Burger model, a viscoelastic solid that has no free flow will result. This represents cross-linked polymers that deform without flow or melting. To summarize, there appear five viscoelastic regions on heating a polymer (according to the Burger model), following the change of relaxation modulus with either time or temperature. We prefer to study the performance of an amorphous polymer, as crystalline ones do not exhibit such sharp transitions. (See Figure 4-17.)

In the figure the five viscoelastic regions are:

Region 1: The glassy state. A constant high modulus (below T_g).
Region 2: Transition.
Region 3: Elastomeric (rubber-like) state. A constant low modulus.
Region 4: Elastomeric flow.
Region 5: Melt flow.

In the case of a cross-linked polymer, only the first three regions are relevant (the dotted line in Figure 4-17).

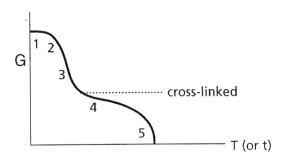

FIGURE 4-17 Time (or temperature) dependence of relaxation modulus

As mentioned, the effects of time or temperature are very similar—suggesting a time–temperature superposition. This enables the use of data from measurements at various temperatures to prepare a master-curve versus extended time, (t/a), where "a" represents the horizontal shift on a logarithmic scale. See Figure 4-18.

A new parameter, the Deborah number, De, is used to define viscoelastic behavior. It is derived by dividing the relaxation (or retardation) time by the duration of the process, as follows:

$$De = \lambda/t \text{ (dimensionless)} \tag{4-14}$$

The name stems from the prophetess Deborah, who said in her famous song "The mountains quaked at the presence of the Lord," that is, even mountains move if one waits long enough. Polymers that have a high molecular weight will also have high Deborah numbers, but after long times this parameter will diminish.

Quite another distinct field of rheology deals with oscillating (periodic) stresses or deformations which lead to dynamic properties. The new prevailing variable is the frequency, ω. One may transform from the domain of frequency to that of time and (by combining, through similarity, the effects of time and temperature) an estimate of properties over a broad range of time is achieved. A regular oscillating stress or strain is periodic, for example, sinusoidal.

$$S^* = S_o \sin (\omega t) \tag{4-15}$$

where
S^* = dynamic stress, and t = time.

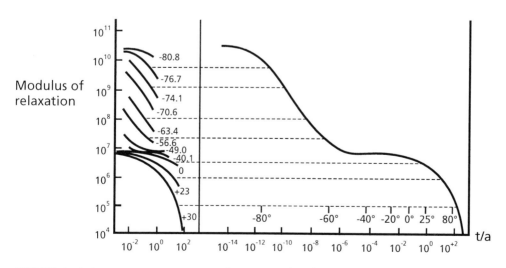

FIGURE 4-18 A master curve (at 25 °C) for a generalized amorphous polymer

The response, a periodic strain, γ^*, may be shifted within the range $0° < \varphi < 90°$.

$$\gamma^* = \gamma_0 \sin (\omega t + \varphi) \tag{4-16}$$

where
φ = the shift angle.

When $\varphi = 0$, both are in the same phase, which is illustrated by an elastic solid. When $\varphi = 90$, the phase shift of 90 degrees represents the behavior of a viscous liquid, wherein the stress is proportional to rate of strain (the time derivative of sin t equals cos t, and is thus shifted by 90 degrees). See Figure 4-19.

It is conventional to express dynamic behavior in a complex plane by combining real and imaginary components:

$$S^*/\gamma^* = G^* = G' + iG'' \tag{4-17}$$

where,
$i = \sqrt{-1}$ (which is imaginary).

Here, the complex modulus G^* is expressed as a combination of a real element G' (representing the in-phase elastic response, the so-called storage modulus), and an imaginary element G'' (representing the dissipated energy, the so-called loss modulus). The ratio of these two moduli is termed the loss factor (tan δ):

$$\tan \delta = G''/G' \tag{4-18}$$

(The loss factor may also characterize thermal transitions, as shown in Section 3.4.)

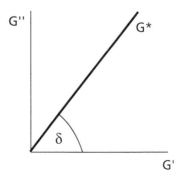

FIGURE 4-19 The complex modulus

The storage modulus G′ varies with frequency in a similar (but reverse) manner to the dependence of relaxation modulus on time (or temperature). Here too it is also possible to locate 5 (or 3) viscoelastic states. (See Figure 4-20.)

There exists a clear relationship between t and $1/\omega$; $G'(1/\omega) \approx G(t)$. Data in a range of $\omega = 10^{-5}$ to 10^{8} is equivalent to that over an extended period of time (thus, avoiding long-term measurements). This is very essential in creep or long-term studies. In addition to G^{*}, other complex parameters may be defined, like compliance or viscosity.

$$J^{*} = 1/G^{*} = \gamma^{*}/S^{*} \tag{4-19}$$

$$\eta^{*} = S^{*}/\dot{\gamma}^{*} \tag{4-20}$$

We have previously introduced viscoelastic mechanical models, but we may also use electronic models as well. The mechanical spring can be replaced by a capacitor, which stores electronic energy, according to Coulomb's law:

$$Q = C\,E \tag{4-21}$$

Here
Q = electrical charge (equivalent to mechanical strain),
E = electrical stress or potential (equivalent to mechanical stress), and
C = capacity (equivalent to compliance).

The mechanical dash-pot can be replaced by an electrical resistor, where Ohm's law

$$E = R\dot{Q} = RI \tag{4-22}$$

replaces Newton's law. Here $I = \dot{Q} = dQ/dt = $ current, and R = resistivity.

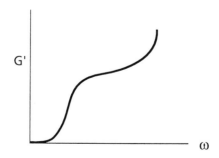

FIGURE 4-20 Frequency dependence of storage modulus

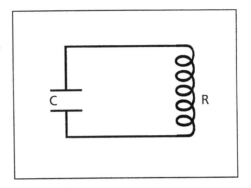

FIGURE 4-21 The electronic Maxwell body

The combined electronic models (Maxwell, Figure 4-21 and Voigt, Figure 4-22) differ visually (from Figures 4-10 and 4-13, respectively). In the first model, the currents are additive (so they are combined in parallel) for a Maxwell model. In the second model, the stresses are added (in series) to construct a Voigt model.

In conclusion, the utilization of artificial models (such as mechanical or electrical analogs simulating viscoelasticity) enables a better understanding of polymer mechanical performance. It also allows a mathematical analysis of the various responses, while frequently a combination of various models or even nonlinear models are called for. A successful combination consists of a modified Burger body. This combines many Voigt elements in series, in addition to a single spring or capacitor and a single dash-pot or resistor, which describe a spectrum of retardation times.

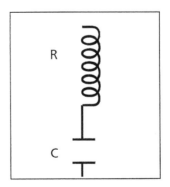

FIGURE 4-22 The electronic Voigt body

4.2 MECHANICAL PROPERTIES OF POLYMERS AND PLASTICS

In the last section we showed the fundamentals of rheology, the theory that interrelates stress and strain, flow and deformation. This section deals with the practical behavior of solid polymers, namely, the mechanical properties. One has initially to differentiate between properties that are measured over short periods and those measured over long periods. Needless to say, the rate of measurement and the temperature dictate the type of performance and the magnitude of resulting data.

Among properties that are measured over short times the most prominent is the stress–strain measurement in tension, but often also in compression or flexure. Several important parameters are obtained during this measurement—tensile strength at break (S_B); ultimate elongation (γ_B); modulus of rigidity/elasticity (E); yield stress (S_y) and elongation at yield (γ_y); as well as the energy to break (a measure of toughness). A typical stress–strain curve is shown in Figure 4-23.

Between A and B, Hooke's law for elastic solids is followed, so that the initial slope stands for the modulus of rigidity, E. The stiffer the material, the higher its modulus, producing a lower elongation at similar stresses. Quite often it is impossible to detect a straight line from the origin and in such case the ratio between stress and strain at a particular standard value is chosen as the modulus (usually at 2% or 100% elongation, depending on the ductility). The point C represents yielding, which is considered the upper limit of elasticity. From this point on, a plastic response prevails, which is truly a flow process (irreversible). Such a response occurs only in ductile bodies, while brittle ones break at the yield point or below. It is only the elastic response that is eventually recovered upon release of the tensile forces. In the region C–D, a decrease of stress is observed, caused by the formation of a "neck" in the specimen, wherein the cross-section drops suddenly in the center while necking proceeds until an even cross section is obtained, as in Figure 4-26.

Elongation is measured by the distance between two control points in the

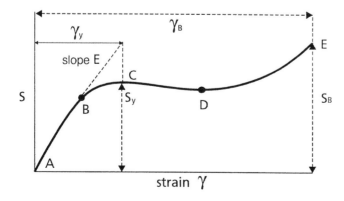

FIGURE 4-23 A stress–strain curve

narrow part of the test specimen (a dumbbell). The final value is affected by the size of the specimen, showing a larger relative elongation when the initial control length is smaller. The stress is obtained by dividing the tensile force by the original cross-section, A_o. This stress is called the nominal stress, S_o.

$$S_o = F/A_o \tag{4-23}$$

However, if the real (smaller) cross-section after stretching is considered (A), the resultant true stress, S, will actually occur.

$$S = F/A \tag{4-24}$$

The relationship between those two stresses is:

$$S = S_o A_o/A \tag{4-25}$$

Whenever the material retains its volume during stretching (incompressibility may occur with ductile or rubber-like materials), one derives a simplified correlation:

$$A_o L_o = A L \tag{4-26}$$

where L = length of specimen and L_o = original length of specimen, so that

$$S = S_o L/L_o = S_o (1 + \gamma) \tag{4-27}$$

The very last portion of the stress–strain curve indicates strain hardening, induced mainly by further chain orientation. In semi-crystalline polymers, one detects an increase in crystallinity and orientation—both leading to an increase of stress. In these cases, $S_B > S_y$.

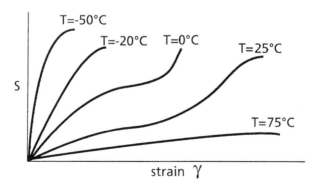

FIGURE 4-24 Stress–strain curves at various temperatures

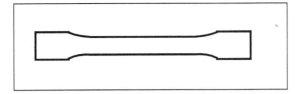

FIGURE 4-25 A stress testing dumbbell

It is obvious that each material exhibits a different performance, while the crucial effects of time (rate) and temperature should not be disregarded. This will be illustrated by a typical schematic plot of a stress–strain relationship at various temperatures. See Figure 4-24. The polymer of choice is LDPE (low density polyethylene). On raising the temperature the elongation increases, while the tensile strength and modulus diminish. Gradually the body converts from rigid and brittle (at low temperature) to soft and ductile. Again the correspondence between temperature and time prevails, so that the type of response at low temperature is identical to that in a short time (high speed).

Therefore, it is essential to test at a standard and fixed speed and temperature for each material. In addition, the specimens must be prepared under controlled conditions, particularly when dealing with crystalline polymers. The area under the stress–strain curve represents the toughness of the material (its ability to absorb impact), expressed by the dimensions of energy per unit volume $\int S d\gamma$.

A typical dumbbell and its necking are shown in Figures 4-25 and 4-26.

As the tensile speed is increased, the toughness data conform with those data obtained directly by the impact strength measurement (which will be discussed later). Definition of some typical materials are illustrated.

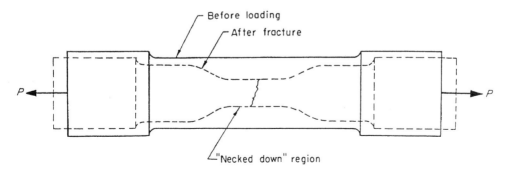

FIGURE 4-26 Necking of a dumbbell

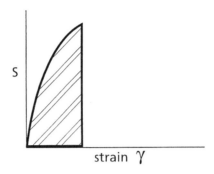

**FIGURE 4-27 Stress–strain graph of
a rigid and brittle material**

Characteristics of Typical Mechanical Properties

Rigid and Brittle   A high modulus of rigidity, low elongation, breaks prior to yield. A small area below the curve indicates low toughness. Typical polymers that give such a response include polystyrene (unmodified) and most thermosets (unreinforced). Brittleness occurs due to low elongation to break (mostly below 2% to 5%). (See Figure 4-27.)

Rigid and Strong   In this case, modulus and tensile strength are high, while elongation is medium, exceeding yield. A typical polymer is rigid PVC. (See Figure 4-28.)

Rigid and Tough   A high modulus, reasonable elongation and strength, and a high energy to break is the case for typical engineering polymers such as Nylon or polycarbonate. (See Figure 4-29.)

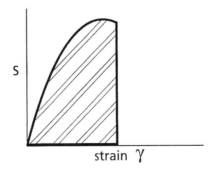

**FIGURE 4-28 Stress–strain graph of
a rigid and strong material**

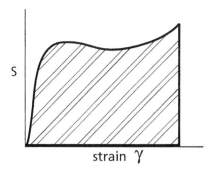

FIGURE 4-29 Stress–strain graph of
a rigid and tough material

Soft and Tough   Low modulus but exhibits high elongation and energy to break. Typical polymers are polyethylene or flexible (plasticized) PVC. (See Figure 4-30.)

Elastomers, on the other hand, have no yield point and therefore no necking, but their elongation and toughness are significantly high. The deformation is basically elastic and therefore also recoverable, as shown in Figure 4.31.

Another factor that dictates the mechanical performance of polymers is the appearance of chain orientation. In the case of fibers, it is customary to pre-stretch in order to achieve a high degree of orientation in the axial direction. In films or other profiles made by extrusion or injection molding, chain orientation is usually obtained in the flow (machine) direction. This is manifested by an increase in tensile strength and a decrease in elongation in the machine direction, while the opposite is true in the transverse direction. This calls for measurements in both directions. In two-dimensional objects this phenomenon is undesirable, as it leads to asymmetric properties (anisotropy). Similar behavior may be found in reinforced materials (with long fibers), in

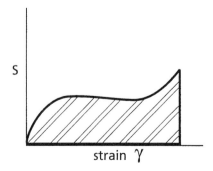

FIGURE 4-30 Stress–strain graph of
a soft and tough material

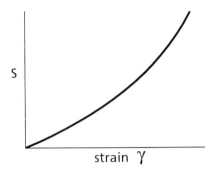

FIGURE 4-31 Performance of elastomers

which fiber orientation occurs. If uni-directional orientation is undesirable, one may use cloth or randomly laid fibers. However, two-dimensional orientation is becoming useful for obtaining improved properties, like bi-axial stretching in blow molding of containers for soft drinks, or the production of thermal-shrinking items that preserve the so-called elastic memory (shrink packages). The effect of orientation is illustrated in Figure 4-32.

The sensitivity to temperature is paramount in the case of thermoplastics, but less important with thermosets. In order to retain strength and toughness (over a broad temperature range) with thermoplastics, one has to shift to more rigid chemical structures, to higher crystallinity or partial cross-linking. This is the domain of those unique materials called "engineering" polymers.

Most polymers illustrate a rather low tensile strength (as compared to metals) in the range of 10–100 MPa. However, strength may be augmented by reinforcement. Epoxy reinforced by glass fibers may reach a tensile strength of 200 MPa, but steel has a tensile strength two times higher (400 MPa). It is illuminating to define the specific tensile strength (per unit density), in which case the strength on the basis of mass becomes quite significant for polymers (due to their low densities), as shown in Table 4-1.

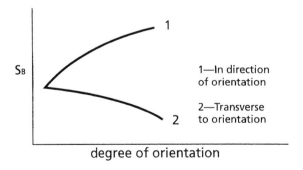

FIGURE 4-32 Effect of orientation on tensile strength

TABLE 4-1
Relative Specific Tensile Strength

Material	Tensile Strength, MPa	Specific Mass	Specific Strength
Polyethylene	13	0.92	14
Nylon	100	1.14	88
Reinforced epoxy	200	2.0	100
Steel	400	7.8	52

The practical conclusion is to increase the mass of polymers, such as using thicker walls in pipes or containers. The ideal is to derive new engineering polymers that exhibit high strength even if not reinforced. The reinforced polymers (namely, the composite materials) excel in strength and stiffness.

The yield strength has been hardly discussed, but is of increasing recent importance. It has already been said that brittle materials break down below the yield stress, S_y, but that ductile materials undergo plastic deformation up to the break stress, S_B. At high temperatures (ductile) $S_B > S_y$, but as the temperature drops $S_y > S_B$ (brittle stage), due to the rapid increase of S_y upon cooling. At a temperature where $S_B = S_y$ we have a transition between ductile and brittle behavior. For semi-crystalline polymers this transition occurs in the range of T_g. It is often important to increase the gap between yield and break strength by either increasing S_B or decreasing S_y.

The ultimate elongation of various polymers lies in a very broad range — less than 1% for reinforced thermosets, up to 500% in the case of polyolefins and around 1000% for elastomers. The use of specimens differing in shape and dimensions affect the results. For materials that have higher elongations, a higher speed of stretching is recommended — for testing thermosets, tensile rate is 0.05 inch/min; up to 20% elongation is 0.5 inch/min, 20 to 100% elongation is 2.0 inch/min, >100% elongation is 20.0 inch/min. Elongation serves as a measure of polymer flexibility and is related (among other factors) to molecular weight. Upon aging of the material, the elongation diminishes significantly. Therefore, elongation at break frequently serves as the criterion for environmental endurance.

The modulus of rigidity expresses the polymer stiffness, and also lies in a broad range. It is less affected (compared to strength) by either speed of measurement or by molecular weight, so that it is truly associated with primary and secondary chemical bonds. With stiff polymers (E > 700 MPa) the modulus of elasticity in tension may reach values of 10^3–10^4 MPa, mainly by increasing the rigidity of the chemical structure or by fiber reinforcement. In the case of soft polymers, E < 700 MPa. With respect to stiffness, polymers are by and large below respective values for various metals (E = 5×10^4 to 2×10^5 MPa). However, systems comprising reinforced engineering polymers can close this gap.

As mentioned above, engineering data regarding strength and stiffness may also be obtained by measuring stress–strain behavior in compression or flexure. Compression testing is mostly used for foamed polymers while flexure prevails for thermosets, exhibiting brittle fracture. Glassy polymers like polystyrene (or glass itself) are stronger in compression than in tension. In addition, there may be a significant characteristic change, like the transition from brittle to ductile response, on switching from tension to compression (including the occurrence of a yield stress). One has to remember that fracture in tension is affected by crack propagation. In compression, flaws tend to close and stop propagation. It is also advisable to calculate the modulus of rigidity of glasses by compression or flexure measurements. In the case of the latter, part of the body is under compressive stress, while another part is under tensile stress. In flexure, it is customary to load until failure or up to 5% strain. Flexure test is shown in Figure 4-33.

The relationships for flexure are as follows:

$$S_{max} = 3P\ell/2bd^2 \qquad\qquad (4\text{-}28)$$

$$\gamma_{max} = 3d\delta/\ell^2 \qquad\qquad (4\text{-}29)$$

$$E = P\ell^3/4bd^3\delta \qquad\qquad (4\text{-}30)$$

where
ℓ = length between supports,
b = width of specimen,
d = thickness, and
δ = deformation at the center under load P.

(Note: Doubling thickness decreases the deformation by eight-fold.)

In practice, it is essential to match the type of measurement with the actual performance of the material (tension, flexure or compression), in spite of the popularity of the tensile test. Sometimes, it is desirable to follow the response to shear stresses. The relationship between the modulus of elasticity E (Young's) and the shear modulus G, is given by:

$$G = E/2 (1 + \nu) \qquad\qquad (4\text{-}31)$$

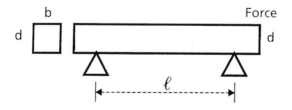

FIGURE 4-33 Flexure test

Here ν represents the Poisson ratio, defined as the ratio between the linear contraction and the elongation in the axis of stretching. In the case of constant volume (an incompressible body like rubber), $\nu = 0.5$ and therefore $E = 3G$; for rigid materials $\nu < 0.3$. There is also a direct test for tear strength, mainly in the case of thin films, similar to those used in the paper industry.

One practical implementation of tensile strength is in designing the thickness of objects. In calculating the wall thickness (t) of a pipe of diameter D, the relationship between the hydrostatic pressure P of the fluid that flows through the pipe and the hoop tensile stress of the material, S, is given as follows:

$$P = 2\,S\,t/(D - t) \tag{4-32}$$

In this expression, the constant 2 serves as a security factor (one may use higher values). It is obvious that t, the thickness, must increase with a rise of hydrostatic pressure. On the other hand, under similar exterior conditions, a larger tensile strength enables the use of a lower wall thickness. Since the extended use of materials is desirable, one can estimate the time-dependent decrease of the strength S, by extrapolating data from measurements at 10^3–10^4 hours to a life span of 50 years. See Figure 4-34. Obviously this extrapolation (using creep tests), may not necessarily be linear, and that is one difficulty in predicting lifetime. However, since there is no better way of evaluation, one should use security factors with caution.

Impact Strength

Associated with toughness, the impact strength is relevant to the endurance of polymeric objects that are susceptible to blows. The direct measurement is based on a hammer pendulum blow to a specimen, frequently center-notched to enhance sensitivity, as shown in Figure 4-35.

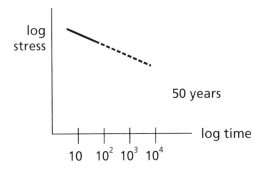

FIGURE 4-34 Decrease of tensile strength with time

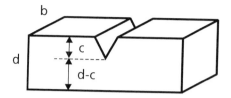

FIGURE 4-35 A specimen for impact test

This method is particularly suited for brittle objects. In the Izod system, the vertical specimen is held at the bottom, while in the Charpy system, a horizontal specimen is held at both sides. The impact strength is expressed in units of energy to break divided by b, the width of the notch, while it may also be divided by the area b (d–c), where c is the depth of the notch, and d is the thickness of the specimen. However, a large portion of the kinetic energy that is measured is dissipated by the vibration of the specimen and the dispersion of the fragments, which should be subtracted from the total energy at break. The occurrence of the notch and the size of the radius of curvature play a decisive role in the test results (including the effect of sample preparation), and one should eliminate an excessively sharp angle in the notch. The notch itself is essential in those cases wherein a ductile fracture turns into a brittle one, which occurs in the range of $S_y < S_B < 3S_y$, and it is calculated that the notch enhances the yield strength by three-fold. On the other hand, if $S_B > 3S_y$ (like in LDPE) the notch becomes useless. It is obvious that notched objects perform differently than unnotched ones.

When dealing with brittle fractures, the Griffith equation correlates the length of the crack ℓ, with the critical stress at break, due to crack propagation as follows:

$$S_B = \sqrt{\frac{2E\gamma}{\pi\ell}}$$

(4-33)

where
γ = surface energy of fracture and
E = modulus of rigidity.

As long as brittle objects are involved, the fracture is caused by structural imperfections that speed up crack propagation. Increasing the thickness of the specimen, enlarges the chances of material flaws. In thin elements, residual compressive stresses may be frozen in (due to a thermal gradient during molding), and in that case, the impact strength of glassy polymers in the notch direction will actually increase.

As stated before, the impact strength is associated with the property of toughness (in spite of the arbitrary method of testing) that is susceptible to too many external factors. As with the tensile test, the direction of the measurement and the degree of crystallinity play an important role. On increasing the amount of crystallites or their size—the toughness of the material diminishes. On the other hand, increasing molecular weight does enhance the impact strength. A relatively new and accurate test method involves rapid stretching by a pendulum up to fracture, namely, a tensile impact test. Novel equipment can measure the history of deformation under an impact stress. Another method, sometimes used for measuring impact strength (mainly for thin films), consists of dropping darts from a standard height to fracture the sample. By varying the weight or the height, the energy of fracture of 50% of the specimens is calculated. The major advantage of this test is its independence of orientation. The normal range for the impact strength of polymers is 1 to 70 kg-cm per cm of notch. Thermosets and rigid thermoplastics exhibit a relatively low impact strength, while soft polymers (like polyethylene) or elastomers verify a high impact strength. An exceptional polymer, polycarbonate, excels both in rigidity and toughness. In general, toughness increases with temperature, mainly above T_g. Among the commodities, polystyrene appears to be a rather brittle material. This has been partly overcome by adding a soft phase of rubber spheres that block craze propagation. Today, high-impact polystyrene (HIPS) is produced by copolymerization with butadiene, leading to a grafted semi-cross-linked structure, which stabilizes the system. (The size of the rubber spheres plays a significant role in the material performance.) In other cases (like PVC) impact modifiers are blended into the matrix. As for thermosets, in order to eliminate the brittleness, various fillers (mostly fibers) are frequently utilized such as paper, sawdust, cotton, or asbestos. Glass fibers may enhance or diminish the impact strength which depends on the polymer type. It is important to note that by applying a bi-axial orientation, a rigid and brittle material (like PS) can become ductile and tough.

Hardness, Abrasion and Friction

Hardness is defined as the property expressing the resistance of a material surface to the penetration of a steel ball (or diamond) under a prescribed load and time. Sometimes it may also be determined by scratching the surface with a standard needle. In the case of rigid polymers, there appears to be a distinct correlation between the degree of hardness and the modulus of elasticity. There are many methods of measurement, which do not show fair consistency. Among the most useful is the Brinell method in which hardness is expressed as the ratio between the applied force (in kg) and the penetration area (mm^2). For most polymers the Brinell indices are in the range of 5 to 50, as compared to 650 for steel. Other practical methods consist of Rockwell or Shore, where each method comprises various ranges of the degree of hardness, differing in geometry and load.

Abrasion is measured by the depletion in mass of a body after being in contact with an abrasive rotating surface, as in the Taber method. Abrasion differs from soft to hard polymers and increases with the coefficient of friction. In principle, it is possible to correlate the resistance to abrasion with hardness for ductile polymers, where ductile polymers have a higher abrasion resistance. Side effects, such as oxidation or chemical degradation which occur during abrasion measurements, may increase the apparent abrasion.

The coefficient of friction characterizes the resistance to mobility between two areas. It is defined as the ratio between the tangential force and the normal force when one body moves against the other. One should distinguish between self friction and friction against a foreign body in addition to dealing with static or dynamic friction coefficients. In essence, a low friction coefficient is preferred for use in bearings, gears or sliding surfaces. Typical friction coefficients for Teflon and Nylon are 0.04 and 0.3, respectively. On the other hand, tires, shoe-soles or flooring necessitate a high friction coefficient, thus avoiding slippage (for rubber, 0.3–2.5; plasticized PVC, 0.4–0.9). During polymer processing, the friction coefficient plays an important role in the solid transportation zone, whereby friction may be reduced by adding lubricants. Friction is associated with hardness and surface structure, but results mainly from forces of attraction and adhesion between two boundaries (therefore, it is larger between identical boundaries than different ones). Temperature and speed of sliding have a remarkable effect on the measured coefficients.

Up to now we have dealt with large stresses applied for short periods. Let's also consider material fracture under conditions of low stresses for long duration—fracture in creep, fatigue, or environmental stress cracking (ESC). Creep under a static load was already discussed for the deformation of viscoelastic bodies, as shown in Figure 4-36.

Upon increasing the load, higher deformations and rates are visualized. There appears to be rapid primary creep followed by a slower secondary one. In the case of thermoplastics, temperature strongly affects the amount of creep. Soft bodies exhibit high deformation, but stiff ones fracture under relatively low loads. (See the fracture envelope.) Creep data (change of modulus and fracture conditions) over a wide range of time and temperature are crucial for engineering design. It is possible to describe, in a three-dimensional diagram, changes in the three parameters—stress, strain, and time. See Figure 4-37.

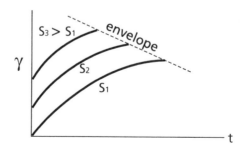

FIGURE 4-36 Creep tests

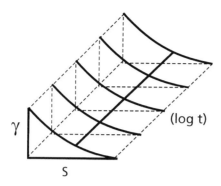

**FIGURE 4-37 Creep and stress
relaxation profiles (three-dimensional)**

Creep appears in the γ–t plane (iso stress), while the stress–strain relationship (at a fixed time) is described by the S–γ plane. Stress relaxation (at constant strain) is described by the S–t plane. Premium engineering polymers or reinforced thermosets illustrate a relatively low creep. Fiber reinforcement usually decreases the creep.

Stress relaxation, another viscoelastic feature, may be expressed for the simple case (a Maxwell model) by a decay function, as mentioned in Section 4.1:

$$S = S_o e^{-t/\lambda} \tag{4-11}$$

where
λ = relaxation time.

While stress relaxation does not control material fracture, it is frequently measured while following the environmental resistance of elastomers. Thermosets show very low relaxation.

Fatigue is expressed as the fracture after cyclic loading during a significant period, when the body is loaded below its break strength. Upon decreasing the load, time to fracture increases. There appears to be an endurance limit, below which the material will not break down (about 20%–35% of the static strength). It is customary to test up to 10^7 cycles. Fatigue test is shown in Figure 4-38.

The resistance to fatigue usually increases with toughness, while occurrence of cracks (or *crazes*) enhances the danger of a fatigue fracture. A side effect which may appear during a cyclic test is a warming of the material (as manifested by hysteresis on the cyclic stress–strain curve), which eventually causes premature failure. Cyclic loading of the material provides general information on performance under dynamic conditions, yielding engineering parameters that characterize the viscoelastic response at various frequencies.

The phenomenon of ESC is typical of polyethylene. It is a combination of

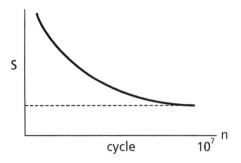

FIGURE 4-38 Fatigue test

loading in the presence of an active medium (mostly polar liquids like alcohol or aqueous solutions of detergents) that does not serve as a solvent. Under the influence of the polar liquid, the surface energy drops and the sample can craze during an ESC test—leading to fracture under a stress lower than the break strength. Under these conditions the fracture becomes brittle. Test specimens are sheets that are folded in order to produce a constant strain and center-notched, immersed in a standard detergent solution in a test-tube at a high temperature (50°C) in order to accelerate fracture. Time to initial crazing serves as a parameter for this phenomenon. (This method is not too reliable as yet.) Increasing crystallinity decreases ESC resistance (when tested at constant strain), so that HDPE appears to be worse than LDPE. (Quite the contrary appears at a constant load.) Increasing the molecular chain length improves the ESC resistance, as well as the presence of narrow distribution, cross-linking, and orientation, or an appropriate copolymerization.

A similar phenomenon (but different in mechanism) is the crazing of glassy polymers (like polystyrene or acrylics) under the influence of various solvents and chemicals, or thermal crazing under load. All these occur with time. It is customary in these cases to refer to a critical strain that may serve as a criterion for the appearance of crazing that leads to brittle fracture of the body. In order to minimize the appearance of crazing, it is important to reduce internal stress (mostly by relaxation), and to avoid mechanical or thermal stress during finishing fabrication such as the use of sharp angles, and so forth.

4.3   OPTICAL, ELECTRICAL, THERMAL AND CHEMICAL PROPERTIES

Optical Properties

After extensively reviewing the mechanical properties of polymers, let's refer to other, less general, properties that are particularly important in cases where these specific properties are dominant. Among the optical properties, transparency (meaning the transmission of visible light) is most important. Transparency is essential for use in construction when seeking a substitute for glass (windows, partitions and ceilings, caps for airplanes, coverage of tunnels and greenhouses in agriculture), in optical appliances (lenses including contact glasses and special optical instruments), in addition to packaging and household use. The most transparent polymers are amorphous ones such as polymethylmethacrylate (better known as acrylic or Perspex), polycarbonate, polystyrene and PVC (with a range of 88%–92% in apparent light transmittance). Transmittance also depends on thickness, so that many polymers transmit light as thin films. Among thermosets such as unsaturated polyester, reinforcement interferes with light transmittance.

Transparency, in general drops with crystallinity (e.g. polyethylene), and with an increase of crystallite size which causes light scattering. Most fillers, colorants and auxiliary additives lead to opacity. Transmittance depends on the refractive index, so that some fillers may preserve full or partial transparency (translucent). There are also dyes that dissolve in the polymer, so that a colored transparent polymer is produced. It is also possible to find stabilizers (including antioxidants or UV absorbers) that do not affect the polymer transparency. Any chemical change in the polymer like degradation or oxidation, or diffusion of some components, may reduce light transmission.

Another annoying factor is the appearance of scratches or flaws on the surface as a result of exposure. In practice light transmission data are needed in a broad range of the spectrum, including infrared and ultraviolet. When a polymer transmits well in the infrared region, such as in covered greenhouses, the soil loses energy on bright and cool nights, as a result of light reflection. Polyethylene transmits 77% of the light in the infrared region (2.5–70 microns) as compared to 12% in the case of PVC. When a polymer absorbs light in the ultraviolet region, it may breakdown due to the high energy absorbed which is of the same order of magnitude as that of the strength of a chemical bond. In order to overcome this deficiency, stabilizers acting as ultraviolet absorbers (UVA) are recommended to neutralize the energy. (Details will be given in Chapter 5.) The acrylics excel in ultraviolet stability, being transparent to UV radiation, but underlying objects should be protected.

Another optical property is gloss, which is determined by light reflection. Most polymers have smooth surfaces and high gloss. However, it is sometimes desired to reduce the gloss by including special additives, like fibers. By and large, gloss is a beneficial property, mainly in household goods, transportation or toys. At present, the property of light reflection is also crucial for the storage of solar energy.

Another optical property is defined as haze, which is of importance mainly in packaging. It is measured by light diffraction, namely as the fraction of impinging light that is diffracted above 2.5°. When haze exceeds 30%, the material becomes translucent.

Birefringence (double reflection) is observed mainly in oriented polymers.

It is measured as the difference or ratio of light refraction in two orthogonal directions.

Electrical Properties

The electrical properties of polymers are of considerable importance in various applications—the insulation of electrical or telecommunication cables, electrical components, electrical appliances and accessories, printed circuits, radar and electronics. Semiconductors based on polymers should also be mentioned. It is also possible to use electrical properties for monitoring or tests without fracture on structure or properties such as sequence of polymerization, degradation, and transition temperatures. Let's discuss briefly some major electrical properties.

Specific Volume Resistivity   Since polymers are covalently bonded, they are, in principle, electrical insulators, due to an absence of free and mobile electrons or ions. The parameter that represents specific volume resistivity (inverse conductivity) is defined as the electrical resistivity, ρ, of a cube having a volume of unity, measured in units of ohm cm. The values obtained for most common polymers lie in the high range of 10^{15}–10^{17}, while polystyrene demonstrates a particularly high value of 10^{19}. The existence of polarity or a double bond (Nylon^{TM} or polyester) reduces resistivity. If conjugated double bonds occur, even lower resistivities in the range of 10^{3}–10^{12} ohm cm are observed, which happens to be in the range of semiconductors. An increase in temperature significantly diminishes the resistivity, which follows an exponential equation,

$$\rho = \rho_0 \exp{(EC/RT)} \tag{4-34}$$

where
ρ = resistivity at T;
ρ_0 = resistivity at reference temperature, T_0;
E = energy of activation;
C = capacity, and
R = gas constant.

Here again, a change of slope signifies the glass transition temperature, T_g. The increase in resistivity (drop of conductivity) depends on a decrease of chain mobility, or a rise of internal viscosity. Consequently, the resistivity rises with increasing chain-length (it may serve as a monitor during the polymerization process), with increasing degree of cross-linking as well as with an increase of crystallinity (or density). On the other hand, the electrical resistivity drops in the presence of ions in the polymer itself (polyelectrolytes) or conductive contamination or residues (including monomers) from the polymerization process (typical in emulsion polymerization). The incorporation of various fillers and other additives (including high-impact elastomeric modifiers or plasticizers, as well as the accumulation of moisture in water-absorbing polymers) all cause a drop of resistivity. The oxidation of the surface, due to exposure to sun or artificial radiation, leads to similar results.

It is therefore essential to identify polymers that do not change their electrical properties within a broad range of temperatures. When electrical conductivity turns out to be desirable in a polymer, in some cases this can be obtained by using conductive fillers (carbon black or metallic powders). In these cases, the resistivity rises with temperature, while a sharp increase occurs around transition temperatures, such as T_m. This phenomenon may be utilized in novel switching and control devices (PTC – positive temperature coefficient). Polyacetylene serves as an example of a conductive polymer.

Dielectric Constant and Loss Factor These properties are important in the use and characterization of polymers. The dielectric constant, ε, is defined as the ratio of the electrical capacity, C, of a capacitor made of parallel plates, when a polymeric insulator is used compared to air or vacuum. Thus, it signifies the stored electrical energy in the capacitor. For most polymers, dielectric constants (at 50 Hertz frequency) appear in a range of 2–6, the values increasing with polarity and with humidity, while dropping with increasing temperature or frequency. The dependence on frequency stems from the fact that some finite time is needed for the charge to respond to a variation of electrical field. With a rise of frequency the allocated response time is diminished, thus decreasing the capacity. Therefore, the polymer response is time-dependent, and its relaxation is highly dependent on structure and viscosity. A complex dielectric constant in an alternating current ε^* may be defined as follows:

$$\varepsilon^* = \varepsilon' - i\varepsilon'' \tag{4-35}$$

The stored electrical energy is related to ε' (the real component of the dielectric constant), while the dissipated energy is related to ε'' (the imaginary component). See Figure 4-39. The loss factor is expressed as tan δ, where

$$\tan \delta = \varepsilon''/\varepsilon' \tag{4-36}$$

The loss is converted to heat (dissipation). For low values of δ, the losses are low so that in the limit $\delta = 0$, there is no loss (this is the ideal capacitor). Low values of tan δ may be approximated as $\tan \delta \approx \sin \delta \approx \cos \theta$, where cos θ is defined as the power factor, θ represents the angle between the direction of the voltage and current in an alternating current. Then Power = Voltage × Current × Power Factor. The loss factor approximately equals the product of the power factor and the dielectric constant, all values taken at a frequency of 60 Hertz. In nonpolar polymers (Teflon, polyolefins, polystyrene)

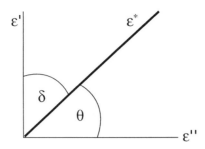

FIGURE 4-39 Complex dielectric constant in alternating current

there appears an extremely low loss factor, 0.0003–0.0009, resulting in a lack of heating up of the object at high frequencies. On the other hand, in polar polymers (PVC, acrylics) the loss factor increases up to 0.05. Such a high loss factor may be utilized for dielectric heating or welding at high frequencies. In this case, the object serving as the insulator of a capacitor heats up in an even and efficient way, compared to heating by conduction.

Dielectric Strength Dielectric strength is characterized by the largest voltage per unit thickness that an object can bear prior to breakdown via a spark or fracture, expressed as kilovolts per cm of thickness. Most polymers have a dielectric strength in the vicinity of 200 kV per cm, while PVC approaches 500 kV per cm. There is a distinct analogy to mechanical strength, so that whatever weakens the object (like existing stresses or crazing) will reduce the dielectric strength. This also relates to a rise of temperature or frequency. Thin objects also display higher dielectric strength as compared to thick ones (with structural defects). This property plays an important role in electrical insulators or instruments which have to withstand high voltage.

Thermal Properties

Transition temperatures that characterize the structure and behavior of polymers have already been dealt with some length. From a practical point of view, limiting temperatures for use is also of interest. One should differentiate between a statistical value derived from use data without material damage and a standard test under prescribed conditions, namely, heat distortion or deflection. In the latter, the temperature is measured wherein the samples undergo a definite deformation under a defined load (usually 264 psi). This temperature is taken to be an upper limit for use of the material without the danger of warping. This value obviously depends on the load (inversely affected). Thermal endurance can also be expressed by time and temperature data that affect mechanical and electrical properties. Data verify that for most polymers, the upper limiting useful temperature is rather low (60°–85°C), but is higher for thermosets (100°–150°C). There are, however, special poly-mers that excel in endurance at high temperatures (HT polymers). In this group, one finds Teflon and silicones (representing older polymers) and a host of novel polymers like polyimides, polysulfone, polyphenyl oxide and aromatic polyamides. Thermoplastics tend to soften as the temperature rises, but the useful temperature may be boosted by using various fillers and re-inforcements, and by cross-linking (chemically or by radiation). A rigid chemical structure (including crystallinity) which elevates the glass transi-tion temperature also increases the use temperature. On the other hand, thermosets do not soften, but often undergo chemical degradation at higher temperatures.

In polymers, heat transfer coefficients in conduction are rather low (K = 0.1–2.5 kCal/hr m°C), so that they are definitely considered as thermal insu-lators (compared to metals where K = 30–300). This, however, reduces the heat transfer during processing or shaping a plastic material, thus diminish-ing the efficiency of heating or cooling. However, the property of insulation is important for thermal insulation, with an additional reduction of conduc-tivity by means of foaming. A low thermal conductivity coefficient is respon-sible for the sense of warmth upon touching plastic materials (like door handles) as opposed to metals. In polymers the thermal diffusivity (a popular concept in heat transfer) appears to be rather low and is around 10^{-7} m^2/s.

Coefficient of Thermal Expansion The coefficient of linear thermal expan-sion lies in a range of $(1-15) \times 10^{-5}$ per °C for most polymers, surpassing that of metals $(12 \times 10^{-6}$ per °C). This calls for stringent consideration when engineering with polymers, in particular when dealing with sandwich and heterogeneous systems (with metals). At the same time, contraction of poly-mers during cooling in the mold must be taken into account, with attention paid to thermal history and phase transitions (crystallization). Reinforced ther-mosets have a lower coefficient of expansion compared to thermoplastics.

Chemical Properties

In this discussion we deal with the resistance of polymers to various chemicals, (mainly water, acids, bases or organic solvents) as well as with their endurance after being exposed to climatic conditions or to fire. Most polymers show very low water absorbency, except for Nylon and cellulose derivatives that are sensitive to humidity. Most polymers also withstand mild inorganic chemicals at ambient temperatures. Excelling at this are the fluoro compounds, Noryl, polyimide and polysulfone; while polypropylene, PVC and epoxy are considered fair. Polyester and polycarbonate are sensitive to bases, while Nylon is affected by acids. Detailed tables of data exist, de-scribing the resistance of plastics to many chemicals at specific temperatures. Most thermoplastics have a tendency to dissolve in specific organic solvents, while solubility drops significantly with increasing chain length or crystallinity. For thermosets, one cannot expect solubility, however, the polymer softens as a result of swelling in appropriate solvents, a process which diminishes upon increasing the degree of cross-linking. The relationship between polymer and solvent may often be expressed by a parameter, called the solubility parameter, δ. When values of δ for solvent and polymer are close to each other, there is a good chance of dissolution. This parameter is associated with cohesion forces in the material and therefore depends on structure. Whenever chemical attraction between polymer and solvent prevails, solubility may be expected.

Permeability (P) is another important property, and measures the rate of transfer of gases and vapors through a layer of polymer (mainly a film). It is expressed as $P = SD$, where S = solubility and D = diffusivity. The character of the gas, its chemical affinity to the polymer, the structure of the polymer and its degree of crystallinity — all strongly affect the permeability (which drops with an increase of crystallinity). This property plays a major role in the packaging of food.

Among environmental variables causing damage to plastics, the most crucial one is solar energy in the ultraviolet (UV) range (260–400 nm), which amounts to around 6% of the light spectrum that reaches the earth, due to the fact that the major part is absorbed in higher atmospheres. However, this is in the range of high energy, reaching the bond strength of $C-C$ or $C-H$ (90–100 kCal per mole). A chain reaction of free radicals can be initiated, combined with oxidation (with the oxygen in air), leading to chemical degradation via chain scission and cross-linking. Even for a polymer that transmits and does not absorb UV light, there are always found absorption sites such as oxidized groups, metallic residues and double bonds. At elevated temperatures, the rate of oxidation speeds up. In summary, most polymers are susceptible to exposure to environmental conditions and need appropriate protection like UV absorbers and antioxidants, or masking agents like pigments (the best appears to be carbon black). Excelling in weather resistance are the fluorine compounds, acrylics and polyarylate. Other aging factors are humidity, heat, temperature gradients, wind, and all kinds of contaminants including industrial fumes. It is possible to follow endurance history by using a laboratory simulator, an instrument that serves as a monitor of accelerated weathering, the so-called Weather-O-Meter. It consists of a unit with radiating lamps (mostly xenon or mercury lamps or carbon-arc) and facilities for watering and control of temperature and humidity. It compares the performance of specimens (changes in mechanical, dielectric, chemical, or optical properties, as well as effects on the surface) to their behavior under "natural" conditions. Currently there is much use of an accelerated instrument, named QUV, in which the source of radiation is comprised of a system of fluorescent lamps, while humidity is provided via condensation on the specimen surface, all carefully controlled as to temperature and time. It is quite difficult to derive an accurate correlation between accelerated and natural performance, so that it is essential to use both methods in order to gain reliable information.

Thermal energy may destroy polymers even in the absence of oxygen, but at temperatures higher than ambient. Such temperatures are, however, within the upper range of processing conditions, calling for the use of thermal stabilizers for many polymers. Specific polymers (HT) have been developed that for long periods can withstand temperatures as high as 200°C–250°C. In a special mechanism, called ablation, the outer layer serves as a thermal buffer. It is carbonized at elevated temperature, thus generating a heat insulating layer to an interior structure. Environmental factors affect the performance of composite materials, in which attention should be given to adhesion between the polymer and the reinforcement, as frequently the attack (by water or other agents) occurs at the interface.

Another crucial factor consists of flame resistance. Except for outstanding polymers like fluoro compounds, Noryl, or aromatic polyamides, it is well known that most polymers based on organic compounds burn well. (There have been many attempts to synthesize non-flammable inorganic polymers, but with minor commercial success.) However, there are polymers that are considered to be "self extinguishing" (more or less) such as PVC and others. Other pernicious side effects include smoke generation and toxic or corrosive fumes. Special additives (namely, flame retardants) act mainly to slow down the rate of burning, but nevertheless do not lead to a comprehensive solution. Various mechanisms of flame retardancy are used, such as the evolution of an extinguishing gas, an endothermic reaction, the appearance of radical transition stages or generation of an insulating layer. A popular measure of fire resistance is the LOI (limiting oxygen index) which is determined by the percentage of oxygen in an oxygen–nitrogen mixture required to keep a flame burning for three minutes in a specified chimney. Teflon gives an LOI of 95.0 as compared to polyethylene at 17.4 and PVC at 47.0. The lowest LOI value for ordinary fire retardancy is 28.0. Many other methods for determining fire resistance exist, such as the rate of flame propagation and the time for self-extinguishing. The Underwriters Laboratories (UL) have defined a very stringent evaluation method.

Lastly, some side effects should be mentioned, the most important being the toxicity of plastics due to evolving monomers (like vinyl chloride, VCM) or additives (lead-based stabilizers, pigments, etc.). Bacteriological attack on plastics should also be discussed though in most cases they do not show enzymatic decomposition (except for plasticizers), which is often considered a disadvantage when waste treatment is concerned. (An extensive treatment of plastics and ecology is given in Chapter 7.) Achievements in increasing the sensitivity to enzymatic decomposition have been demonstrated by introducing polysaccharides or other blends into the polymeric system. Another successful solution to waste management is based on harnessing solar radiation (in the UV region) to cause a rapid breakdown (possibly under control) of plastics films and containers (used in agriculture or packaging) by making use of specific accelerators. A search for biodegradable polymers is characteristic of an era of rising ecological awareness.

4.4   STRUCTURE–PROPERTY RELATIONSHIP

Having presented a general discussion of physical and chemical properties of polymers and plastics, it is appropriate to analyze the effect of structure on mechanical properties. The chemical and steric structures determine the strength of primary and secondary bonds, the location of transition temperatures, as well as the morphology. These act in addition to the effects of chain dimensions — molecular weights and their distribution.

Polymer stiffness, as represented by the modulus of elasticity, depends mainly on the chemical structure and the morphology. Rising molecular weights contribute to an improvement of strength (like tensile and impact strengths, ultimate elongation) but have no significant effect on modulus and yield stress. Crystallinity enhances tensile strength (particularly the yield stress) but diminishes elongation and impact strength. (Small spherulites are preferred in order to obtain better mechanical and optical properties, except for deformability.)

Orientation also enhances the mechanical strength and diminishes the elongation (all in the tensile direction), but an increase of S_y may increase brittleness. Cross-linking has in principle a similar effect as crystallinity, and may lead to fracture in extreme cases.

A broad molecular weight distribution may not be desirable for good mechanical properties but it definitely improves workability. Most correlations of mechanical properties and molecular weight refer to the weight-average, $\overline{M}_w$, but other averages (not always well defined) are also recommended. An improvement in performance upon increasing molecular weight (including chemical and climatic resistance) quite often reaches a plateau, beyond which diminishing returns prevail, but difficulties in processing (due to rising viscosities) are enormous. The crystallization process itself may by hampered by an increase in viscosity. In general, the dependence of strength (S) on chain length (DP), may be expressed as follows:

$$S = S^\circ - A/DP \tag{4-37}$$

where
S° = tensile strength at DP $\rightarrow \infty$
A = constant.

Usually no mechanical strength appears below a degree of polymerization (DP) of 30, while asymptotic value is reached around DP = 600. It is important to keep a balance between strength, stiffness and toughness, preserving these properties within a wide range of temperatures. In the case of modulus of rigidity, it is possible to approach the theoretical value in oriented polymers.

On the other hand, a large gap exists between the measured and calculated strength. According to the bond strength of $C-C$, an estimated $S = 15,000$ MPa is derived. Using strength theory, a theoretical value of around $S = E/15$ (where E stands for the Young's modulus) is obtained for brittle objects. However, in reality, the measured strength appears to be 1% to 10% of the calculated value. This discrepancy is due to the existence of weak spots and structural defects. It may be partially overcome by the use of thin and highly oriented fibers, namely whiskers. In view of the strong dependence on history and rate, as well as on temperature and geometry, data obtained for the same materials show a high degree of variability for repeated measurements, resulting in a large spread in numerical values obtained from literature. In addition, there exists a wide range for each polymer, depending on molecular weight and crystallinity, so that rough average values must be contended with. In any case, well specified standard measuring methods are essential.

An improvement in mechanical properties is achieved by modifying the chemical structure, morphology or chain architecture. Stiffening the chemical structure is obtained by using aromatic or heterocyclic rings in the chain backbone (not in side-chains), and by replacing hydrogen–carbon bonds by nitrogen or oxygen bonds. This enhances primary bond strengths, improving also the strength of secondary bonds (polarity). (Extraordinary structures have been preferred such as the so-called ladder, wherein two conjugated bonds have to be taken apart prior to chain breakdown.) The stiffness of the structure improves both mechanical and thermal properties (most prominently in aromatic polyamides and polyimides). Whiskers (as mentioned above) consist of highly crystalline polymers obtained through the pyrolysis of polyarcrylonitrile or polyimides, creating a graphite-like pure carbon structure. The combination of polymer (or elastomers) and reinforcing fibers leads to extremely excellent properties. Table 4-2 summarizes the effects of the structure on ultimate performance.

4.5 SUMMARY

Although there has been an increase in the understanding of structure-property relationships of materials, much is yet unknown, calling for a discrete treatment of data for each individual polymer. By improving measuring methods, the existence of transition temperatures (or relaxations) can be detected over an exceedingly wide range, as well as highly reliable information on molecular structure (molecular weight distribution, branching, etc.), on morphology (crystallinity, tacticity, amorphous structure, orientation), and on the fracture structure.

TABLE 4-2
Polymer Structure–Property Relationship

Property	Molecular Weight	Broad MWD	Crystallinity	Branching	Cross-linking	Polarity	Rigid Chemical Structure
Tensile strength	+	−	+	−	0	+	+
Yield strength	+	+	+	−	+	+	+
Elastic modulus	0	0	+	−	+	+	+
Toughness	+	−	−	+	−	+	−
Hardness	+	−	+	−	+	+	+
Chemical resistance	+	−	+	−	+	−	+
Solubility	−	+	−	+	−	−	−
Softening temperature	+	−	+	−	+	+	+
T_g	+	−	+	−	+	+	+
Density	0	0	+	−	0	+	0

Modern technical data also include many dynamic properties obtained by measurements over a wide spectrum of frequencies and temperatures, in addition to data on mechanical properties in a wide range of times and rates. Quite often the data may be described by a master curve providing extremely widely applicable information with some sacrifice of accuracy. On the other hand in most cases, the material consists not only of a single polymer but of a host of additives and modifications. At present there is much use of copolymers in a variety of compositions (including blocks and grafts), polyblends, composite systems and many stabilizers and additives. In practice, a giant additives industry has been developed along with the resin industry.

In addition, the conventional division between thermoplastics and thermosets is often losing distinction, while hybrid materials (ionomers, and elastomers of the SBS block type) are appearing, which behave as thermoplastics at elevated temperatures and as thermosets (most important for elastomers) at ambient temperature. Moreover, many thermoplastics are now also modified by cross-linking (either by chemical or radiation process), so they can serve dual purposes—good workability as a thermoplastic is combined with the improved thermal and mechanical stability of a thermoset.

PROBLEMS

1. A rheological equation is expressed as

$$\eta_o k \dot{\gamma} = k\tau + (k\tau)^2$$

Derive an equation $\eta = f(\tau)$.
Is it linear with τ?
Rewrite the equation as a linear dependence on shear.
Evaluate η at $\tau = 0$; $\tau = \infty$.
Does this equation have a name?

2. A non Newtonian fluid may be described by the following rheological equation:

$$B\eta = \frac{1 + A\tau^{n+1}}{1 + C\tau^{n+1}}$$

Define explicitly: τ, $\dot{\gamma}$, η_o, η_∞ for this fluid.
If $C > A$, is the fluid shear thinning or shear thickening?

3. A polymer melt behaves according to the following flow equation: $\tau = \dot{\gamma}^{1.5}$.
    a. What type of liquid is it? (Newtonian, shear thinning, shear thickening).
    b. Express the apparent viscosity as a function of shear stress.
    c. Express the apparent viscosity on a straight line.
    d. Will this melt flow better than a Newtonian fluid?

4. A non Newtonian fluid is described by the following equation:

$$\tau = \tau_y + B\dot{\gamma}$$

    a. Plot the flow curve (τ versus $\dot{\gamma}$).
    b. Express the apparent viscosity and plot it versus shear rate.
    c. Is this fluid Newtonian? Comment.
    d. What is the name for such fluids?

5. The following rheological equation describes the flow of a certain polymer melt:

$$\frac{1}{\mu} = \frac{1}{\mu_o} + b\tau^{-\frac{1}{3}}$$

    a. Express μ/μ_o.
    b. Express graphically by a straight line.
    c. Express as a function of shear rate.
    d. Find apparent viscosities at low and high shear stresses.
    e. What type of fluid is it?

6. The following viscoelastic elements are tested for stress relaxation (at constant strain). Which one will show stress relaxation? Comment.
    1. A Kelvin body.
    2. A Burger body.
    3. A Kelvin body combined in series with a spring.
    4. A Kelvin body combined in series with a dash-pot.
    5. A Kelvin body combined in parallel with a dash-pot.

7. Plot the creep and recovery history of the following bodies. What happens in each case in short or long times? (Please, no differential equations!)
    1. A Maxwell body.
    2. A Maxwell body whose modulus is twice as much.
    3. A Maxwell body whose viscosity is twice as much.
    4. A Kelvin body combined in series with a spring.
    5. A Kelvin body combined in series with a dash-pot.

8. Estimate melt viscosity of a polystyrene sample under conditions of low shear stress, at 150°C. You may assume that for the same sample, viscosity at T_g is 10^{13} poise.
    At what temperature will polycarbonate have the same melt viscosity?

9. Write the relaxation equation of a Maxwell body. Derive the following:
    a. Change in stress at $t = \lambda$. What is the Debora number at this moment?
    b. What will be the stress at small or high Debora numbers?
    c. Express by J(t), compliance of relaxation.

10. Two different polymers, A and B, were tested in a tensile machine, at high speed. Polymer A is characterized by a stress–strain curve that follows a

Kelvin model, with modulus of elasticity of 10^8 dyne/cm^2 and retardation time of 10^{-5} sec. Polymer B follows a Maxwell model with modulus of elasticity of 10^{10} dyne/cm^2 and relaxation time of 2 seconds. In both cases the tensile strain rate was 5% per second. Polymer A broke after 40 seconds and B after 30 seconds.

Calculate:
i. The ultimate strain for each polymer.
ii. The tensile strength for each polymer.
iii. Which one is tougher (higher impact strength)?

11. Determine the thickness of a rigid PVC pipe, that has to deliver water at 8 atmospheres of pressure, at 30°C. The internal diameter is 2 inches. Tensile strength of rigid PVC, as measured at 23°C is given as 330 kg/cm^2. A drop of 2% in strength is assumed per each centigrade degree. In addition the strength decreases with aging. At 23°C the tensile strength dropped to 220 kg/cm^2 after 1000 hours.

Assuming a linear correlation (on log : log scale) between tensile strength and time of aging, calculate the desired thickness of the pipe for twenty years of lifetime.

12. Compare mechanical properties of polyethylene, polystyrene and polycarbonate. Can you explain the differences in performance in view of molecular structure? Crystallinity? Thermal transitions? Which one has the best properties?

13. A stress–strain curve of a certain polymer fits the following equation:

$$S = B\gamma - A\gamma^2$$

a. Express constant B as a function of the modulus (initial slope).
b. Express constant A as a function of modulus and strain at yield.
c. Express yield stress.

14. Plot schematically the stress–strain curves for the following polymers. Comment.
a. Ploystyrene at room temperature.
b. Polystyrene at high temperature.
c. Polyethylene at room temperature.
d. Polycarbonate at room temperature.

15. A pipe for domestic water supply is designed for a pressure of 10 atmospheres. The diameter (interior) should be 125 mm and the thickness 10 mm.

It is estimated that the tensile strength after 50 years will drop by 80%.

Which of the following polymers will be suitable?

    a. LDPE

    b. HDPE

    c. PP

    d. PVC

    e. ABS

What is the required thickness for the polymers that do not meet the specification?

What polymer will you choose?

5

Compounding and Processing of Plastics

*T*HE MAJOR GOAL IS to make durable products that are composed of polymers and suitable additives, namely plastic materials. This process is generally composed of two stages—a) preparation of the final compound (compounding) by homogeneous dispersion, and b) forming the final shape (shaping). In this chapter the various problems involved in making the final product will be illuminated.

5.1 ADDITIVES FOR THE PREPARATION OF PLASTIC MATERIALS

Various components are commonly added to the polymeric resin in order to obtain desired properties or preserve material homogeneity throughout the fabrication process and its utility lifetime. Therefore, a precise recipe is prepared while the system is brought into optimal mixing conditions. The various additives to the polymer range from a concentration of fractions of a percent up to tens of a percent. Sometimes compounding is performed at the plant where the polymeric resin is manufactured, but due to the variety in recipes, only basic compounds are usually supplied. On the other hand, there are plants that deal only with compounding including the preparation of master batches. In this case too the final decision regarding the desired composition lies with the producer who fabricates the end product, and deals with the final compounding in his plant. Let's start with a description of various types of additives, and then briefly explain the technology of compounding.

Stabilizers

The various stabilizers are considered the most essential additives, sometimes comprising the only additive present in the system. The addition is in the order of fractions of a percent up to several percent.

Thermal stabilizers eliminate premature decomposition of polymers under the conditions of elevated temperature prevailing during processing. One should remember that most polymers decompose around 250°C–300°C. Due to the presence of oxygen in the environment, specific stabilizers that guard against oxidation (so-called antioxidants) are also used.

A large collection of stabilizers is recommended for different polymers, wherein the most important issue is compatibility between stabilizer and polymer. One of the most susceptible materials to degradation is the useful polymer PVC, where various metallic salts (tin, zinc, calcium, barium) or other compounds such as phosphites in various combinations and in synergistic blends are recommended. (The term "synergism" will be discussed later.) The use of lead compounds to stabilize PVC has been ruled out due to health hazards. Without stabilization at an appropriate concentration, PVC will decompose, generating double bonds (leading to dark colors). It will combine with oxygen from the air, which eventually causes a multistage degradation process composed of depolymerization (chain scission) and cross-linking. During this decomposition, an evolution of HCl will take place, a rather toxic and corrosive material. Only a small number of stabilizers are permitted in the packaging of food or materials that come in contact with human beings. The most highly recommended stabilizer for these uses is comprised of organic tin compounds. In addition to the vinyls, most polyolefins, polystyrene, ABS and others are stabilized.

While the heat stabilizers act directly to eliminate chemical decomposition (sometimes via a mechanism of combination with the decomposition products in order to stop chain reactions), the use of antioxidants is aimed at the elimination of oxidation during processing as well as during lifetime. Some polymers (mainly elastomers) are particularly sensitive to attack by ozone. In essence, the antioxidants serve as oxidation retardants, which act in the presence of a source of energy like heat or light radiation. The chemical process is described by a series of free-radical reactions producing among others hydroperoxides and peroxides—ROOH, ROO*, RO*. The reaction is autocatalytic, so that it should be stopped at an early stage. Active antioxidants are based on several groups—amines or various phenols react with radical peroxides and neutralize them; or phosphites and sulfuric compounds attack hydroperoxides and eliminate the formation of free radicals. The polymers most sensitive to oxidation are those that contain unsaturation like polybutadiene or other elastomers, as well as polystyrene, ABS, Nylon and polyolefins. Within the polyolefins the existence of tertiary carbons increases the sensitivity to oxidation, like in polypropylene or branched polymers (LDPE). The occurrence of a crystalline phase diminishes the rate of absorption of oxygen. The recommended concentration of antioxidants (AO) is in the range of 0.1%–0.3%.

A third type of stabilizer consists of ultraviolet absorbers (UVA) that are crucial for environmental protection, and are essential for elements that are exposed to the atmosphere. While the previously mentioned stabilizers (mainly AO) do part of the job, it is essential in many cases to act directly against the source of light in the UV region, which is the major contributor to the environmental aging of plastics. Here, too, the optimal type of stabilizer (or blend) has to be chosen. The advantage of synergism may often be exploited, to achieve an improved activity by combining different types of stabilizers and surpassing the contribution of the individual additive stabilizers.

Among the stabilizers some act directly as UV light absorbers (mainly in the wavelength of 230–390 nm) such as compounds based on derivatives of hydroxy-benzophenone or benzotriazole. Quite another type of stabilizer acts by a mechanism of energy transformation (neutralizing or quenching high energy). The most prominent consist of complex organo compounds of nickel (chelates). Often stabilizers with two different mechanisms are used together. The adherence to the polymer and elimination of diffusion to the environment are both essential; however, they should be mobile enough to move to the surface layer. Novel stabilizers, based on steric hindered amines (HALS) are considered to be very efficient, albeit there is lack of agreement regarding their stabilization mechanism. It is assumed that there exists a mutual reaction of antioxidation and hydroperoxide decomposition combined with the destruction of free radicals. These stabilizers are essential for protecting polyolefins and other polymers.

Various polymers are more sensitive to specific UV wavelengths in the region of 290–360 nm and the type of UV stabilizer should be appropriately fitted. UVA are heavily utilized in agriculture, construction and transportation with the following polymers among others–polyethylene, polypropylene, PVC, polyester, polycarbonate. The useful concentration is 0.1%–1.0%. There is much interest in the development of reactive groups within the polymer chain for improved weather resistance. There is demand for a stabilizer that will combine protection against oxidation as well as against energy sources like light and heat.

Various pigments act as radiation screens (reflecting light radiation), the most important being carbon black, zinc or titanium oxides. However, these pigments also block light transmission or often impose an unintended color. But when carbon black can be tolerated (about 2% is sufficient), a light protection of tens of years is obtained. (It must be emphasized that carbon black is not only a light reflector but is also effective in eliminating the activity of free radicals and stabilizing against oxidation.)

A fourth type of stabilizer consists of additives against flammability (fire retardants). As mentioned before in Section 4.3, we indicated that, in most cases, it may be possible to retard flammability by making use of appropriate additives (sometimes built-in within the chain). However, it is very difficult to convert a burning polymer to a completely nonburning one. The useful fire retardants include organic compounds with bromine, chlorine or phosphorus, as well as inorganic retardants based on salts of antimony, boron, magnesium and alumina (quite often in a blend with the former group). The inorganic retardants have the advantage of avoiding toxic emissions and smoke. The required concentrations are based on the reactive element. For polyester, as an example, the recommended doses are 12–15% bromine, or 25% chlorine, or 5% phosphorus.

A premium family of fire retardants is based on reactive compounds that are able to enter into the polymer chain during the polymerization process (mainly polyesters, epoxides and polyurethanes), or monomers that contain fire-retardant groups (like bromo-styrene). Due to their relative physical stability (regarding diffusivity to the atmosphere), the advantage over additives is clear. In this case, it is also important to study the effects on mechanical and optical properties and mainly on weather resistance.

Synergistic combinations of several families of fire retardants are frequently preferred (halogens with phosphate compounds, halogens with antimony), as the combination is more effective than the individual additive contribution. Some compounds act in a dual way, like fire retardancy and filling or plasticizing. Much in use are chlorinated paraffins as well as phosphite esters (often also including halogen). Fire resistance is essential in construction (unsaturated polyester, polyurethane foams, polystyrene and PVC), furniture including rugs, transportation and electronics.

Some stabilizers are aimed at very specific uses such as antistatic agents (eliminating static electricity) and stabilizers active against bacteriological attack. Currently there are also additives for controled degradation of the polymer after the product has been used.

Fillers and Reinforcement

Fillers may be used at concentrations of 10% to 50%, targeted at some desired physical or chemical properties, but also frequently useful as cheapeners. Wherever their major utility is to stiffen and strengthen, they will be termed "reinforcement," and in most cases have a fibrous structure. Useful fillers include limestone, quartz, silica, talcum, alumina and other minerals. Particle size and distribution are of highest importance. Low-cost fillers for thermosets (eliminating brittleness) include sawdust, paper or jute. The use of ground limestone (or precipitated $Ca\ CO_3$) is mainly found in PVC and unsaturated polyester in the fields of construction and flooring. Currently, low cost fillers are also added to polyolefins and to most polymers, without reducing the profile or impairing melt flow.

In order to enhance the adhesion of fillers and polymer matrix an interesting industry of coupling agents has been developed, including stearates, silanes and, recently, titanates. These materials show chemical affinity to both polymer and inorganic filler. In general, fillers improve dimensional stability and impact strength, mainly for brittle thermosets.

Asbestos also aids in thermal stability (but its use is limited because it may be carcinogenic) while mica is distinguished in its effects on electrical properties (for insulation). In addition, there are currently fillers that convert a polymer into an electrical conductor (or semiconductor) using mainly carbon black or metallic powder.

A group of fillers is distinguished by high reinforcement; mainly comprising fibers in which the ratio of length to diameter exceeds 100. Among the reinforcing fillers are glass fiber, asbestos, cellulose fiber and paper. In addition, there are a large collection of synthetic fibers based on polymers, like — Nylon, polyester (unsaturated) and, currently, aromatic Nylons (aramides) like Kevlar. High-performance (and expensive) reinforcing fillers include carbon (mainly graphite) fibers, alumina, silicon, carbide, nitrite, boron, beryllium and other metals. Producing extremely high strength and stiffness, these specific fibers have been developed for space, aviation and military use.

Currently most thermosets are reinforced, but many thermoplastics are following suit. Pressed slabs made of reinforced material are called laminates. The combination of polymers and reinforcing fibers serves as a basis for composites, as well as the combination of polymers with ceramics or metals.

For the commodity polymers, various types of glass fibers are still used in abundance, including mainly type E glass. Since this fiber has a tensile strength of about 3 GPa and a modulus that approaches 100 GPa, the large contribution to the strength and rigidity of a reinforced plastic (assuming laws of additivity) is evident. With alumina fibers, the strength and rigidity are ten times those for glass-fiber reinforcement. In most cases, thermal stability, dimensional stability (including reduced creep), as well as impact strength, are all improved. Fibers contribute when the length to diameter ratio is high (continuous fibers) mainly in the longitudinal direction and show anisotropic properties as in oriented bodies. Fibers should be surface treated by coating with coupling agents. When isotropic properties are desired, layers of fiber mat or fabric may be placed in perpendicular directions. When no premium constructional requirements are needed, the use of mats is quite common.

In processing thermoplastics, the use of short fibers is common, losing the advantage of long continuous fibers which can be cut during injection molding or extrusion. Polymer fibers have the additional advantage of light weight. A blend of various fibers may be recommended for specific cases. It is sometimes customary to use nonfibrous geometries (like full or hollow spheres). Carbon black is also utilized as a unique and essential reinforcing filler in natural or synthetic rubbers.

Plasticizers

Plasticizers are frequently incorporated to improve the workability of polymers, but often transform a rigid plastomer into a soft and ductile material. From early times, camphor was used to plasticize nitrocellulose and yield celluloid. Today plasticizers are most common in soft PVC, while unplasticized rigid PVC is also extensively used. Plasticizers may consist of liquids or solids (oils, esters or prepolymers). They are characterized by an extremely low glass-transition temperature, weakening the secondary bond strength in low-crystalline polymers with which they form a solid solution phase.

PVC has strong secondary bonds due to its polarity, which makes it rigid (high modulus) and of high T_g. Upon the incorporation of plasticizers, the elastic modulus and T_g both drop, while the toughness rises — leading to elastomeric properties which are desired in a wide range of utility.

Many plasticizers are based on phthalic (or adipic) esters, the most common in use being dioctyl-phthalate (DOP). A large variety of plasticizers of many structural families is offered including blends. It is essential to have compatibility between polymer and plasticizer, in order to obtain homogeneity. One way of attaining compatibility is by adjusting solubility parameters. Phosphorous compounds function also as fire retardants. The performance of a plasticizer is measured by the intrinsic effect on flexibility (mainly at low temperature). But the total system stability should be considered, involving the elimination of eventual diffusion from the body to the atmosphere or side effects on polymer endurance in weather and fire.

The usual concentration of plasticizers is 20% to 40%, but some systems (plastisols) have 50% to 60% plastification. The diffusion of plasticizers impairs flexibility, making the polymer rigid and brittle. It is important to eliminate the use of toxic additives, which adversely affect the environment through contact with food stock or other polymers. For these reasons a liquid-phase plasticizer may be replaced by a solid-state plasticizer which is mainly based on a short-chain polymer (polyester or epoxy) or a long-branched one. The most ideal plastification may be achieved by an appropriate copolymerization. A typical example is a copolymer made from vinylchloride and vinylacetate, in which a low T_g monomer is inserted into the main chain.

Colorants

Most colorants are pigments which are not soluble in the polymer, producing a colored but opaque product. When soluble in the polymer (dyes), transparent colored objects are obtained. Pigments are most abundant, consisting of a large collection of colorants, mostly inorganic compounds, but organic ones as well. For compatibility with the polymer, the pigment must be stable during processing and use. It should stay dispersed in the system without diffusion or bleeding. An ideal dispersion of thin powders that are uniformly and stably dispersed in a particulate solid polymer, or in a viscous melt, is not an easy accomplishment. It is much easier to use a master batch, prepared by a plant with expertise in compounding (at a pigment concentration of 35% to 50%), and then dilute it during processing, as the achievement of a coherent and identical color is often a real art. Among the most useful pigments are titanium dioxide (representing the most efficient white pigment), various iron oxides that appear in many colors (red, black, yellow and brown), cadmium and various chromates. There are special pigments like metal powders (aluminum, and others) as well as pigments for special effects. There are many families of organic pigments which will not be described here. Carbon black belongs to this category and is considered the most commonly used black pigment (with additional properties such as weather-resistance, reinforcement and stabilization against static electricity). There are many types of carbon black, varying by mode of preparation and particle size. It is recommended to use a carbon black with a large surface area (small particles).

Special Additives for the Improvement of Workability and Performance

Here we will mention leftovers–lubricants that diminish the friction between polymer particles (internal) or with the walls of the fabricating machines (external). An internal lubricant affects mainly a reduction of viscosity while an external one affects the surface layer and augments the output (provided no slip occurs). Many lubricants serve simultaneously as internal and external agents, as a borderline is not that sharp. The polymer most demanding of lubrication is PVC for improving workability and elimination of degradation. However, other polymers are lubricated as well, like polystyrene, polyester and polyolefins. Among the common lubricants are fatty acids and their salts, paraffins, and low MW fractions of polyethylene. Only a small concentration should be used in order not to affect the ultimate properties. In the case of broadly distributed PE, self lubrication occurs.

Another treatment of the surface to improve workability is based on coating the mold with a layer that eliminates sticking (like silicon); namely, release agents. Other accessories that may appear in a compound are surface active agents (wetting the fillers and pigments); impact modifiers; blocking agents, slip agents, coupling agents and blowing agents (as solids, liquids or gases); catalysts; hardeners and sometimes retardants (for thermosets); and cross-linking agents (mostly peroxides). We have not mentioned (while discussing compounding) the growing use of inter or intra polymeric mixtures, namely, polyblends. The results are very promising, as long as compatibility is achieved–leading to a homogeneous phase. On the other hand, it is sometimes desirable to create a separate phase for special needs (as for impact modification), but if possible, copolymerization is advantageous. The polyblends have created a revolution in the polymer world. The most prominent polyblends are Noryl (PPO with styrene) and a polyblend between polycarbonate and ABS (Cycoloy). Polyblends or alloys (compatible) are very important engineering polymers, their use reaching 20%. Whenever incompatibility prevails, special coupling agents (so-called compatibilizers) are frequently used. Details on the properties of selected polyblends appear at the end of Chapter 6.

Let us now survey the compounding process itself. As mentioned, some components already appear in the polymer, being introduced by the resin manufacturer (mainly heat stabilizers and antioxidants) but one should always check to see if their concentration is sufficient and whether the compound fits specific requirements. The use of a master batch has been described, wherein the fabricator has to undertake an accurate diluting process.

The mixing of various components is tedious, requiring high energy and long times for a homogeneous compound, often involving the mixing of polymeric solid pellets with various powders or possibly with liquid plasticizers or stabilizers. It is easier to mix when the polymer itself is provided as a powder as in the case of PVC. In any case, it is difficult to be accurate (in concentration) when minute quantities of additives are added. It is possible to mix in stages, first dry-blending with rotating vessels or blades at low speed. In order to enhance the efficiency of mixing, high-speed mixers are used, causing shearing and melting of the polymer. Some equipment has been transferred from the rubber industry (roll mills or internal mixers of the Banbury type) for compounding PVC, polyolefins and others. An internal mixer requires high energy, so short mixing times and minimal overheating are essential for the polymer stability.

The compound leaves the mixer via a die for cutting into pellets through an attached screw or a grinder. The equipment is mainly operated in a batch mode, but continuous equipment is also available, mainly for extruding homogeneous pellets. The advantage of continuous mixers is evident, as higher homogeneity is obtained (as related to variations between batches), and better adaptivity to a continuous production line. Mixing extruders have a different structure than ordinary ones. Much use is also made of twin extruders.

A theoretical analysis of solid mixing is rather complicated. Statistical units are used based on the variance of concentration in different zones of the compound, leading to a parameter, termed the intensity of dispersion. Other parameters based on the dimension of length are connected to the real location of the dispersed particle.

As melting of the polymer most frequently occurs, liquid flow of the continuous phase is considered. In this case, conditions that generate turbulence are important for optimal mixing, but frequently cannot be obtained due to the high viscosity. Quality control regarding compound homogeneity is most desirable. In the extreme case, poor mixing may be visually detected, but specific equipment for controlling an even dispersion is called for to measure concentrations by optical, mechanical or electrical methods.

5.2　EXTRUSION OF THERMOPLASTICS

During the process of extrusion, simultaneous operations occur in order to produce objects continuously—solid transport, melting, compacting and flow under pressure through a die that provides the ultimate shape. Actually, a whole array of units for cooling, stretching and cutting is attached. This comprises the most important continuous processing device directed to handle infinitely long elements—rods, fibers, pipes and profiles, sheets and films, wire coating, and so on. The heart of the extruder is the screw, continuously turning in a barrel, thus performing the necessary transport functions. The screw currently serves also for transport and melting in other fabrication processes, such as pelletizing, injection and blow moldings. In all these processes, it successfully replaces a ram that was once utilized even for extrusion itself. The use of extruders actually started in the rubber industry that preceded the plastics industry (mid-19th century). Since 1925, growing use occurred in the extrusion of thermoplastics followed by machine development. In essence, most thermoplastics can be shaped via extrusion, preferring grades of relatively high molecular weight (high melt viscosity and strength). The theoretical analysis of the process has attained serious achievements since the 1930s while major effort in the mathematical analysis of the extrusion process started in the 1950s. The melt flow was simplified in order to obtain an initial analytic solution that relates the output to the operational variables. With time, more exact (numerical) solutions were developed considering complex liquids (viscoelastic, non-Newtonian) and the geometry of the fabrication process. The operational conditions were converted to computer control.

In this chapter, the principle of the process and structure of the equipment will be discussed. A schematic description and a simplified analysis will be undertaken, leading to an approximate analytical solution. Let's observe the

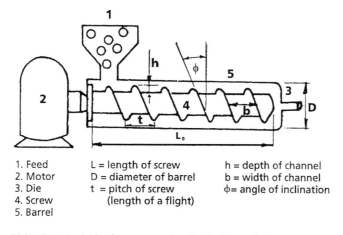

1. Feed	L = length of screw	h = depth of channel
2. Motor	D = diameter of barrel	b = width of channel
3. Die	t = pitch of screw	ϕ = angle of inclination
4. Screw	(length of a flight)	
5. Barrel		

FIGURE 5-1　A single screw extruder (schematic)

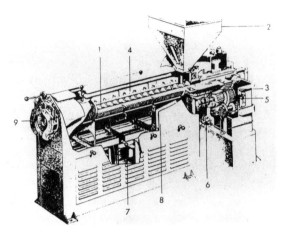

**PHOTO 5-1 Section view of a single screw
plastics extruder**
**1, screw; 2, hopper; 3, feed section; 4, barrel
heaters; 5, gear box; 6, lubrication system; 7, air
blowers to control barrel heating and cooling
temperatures; 8, double walled hood for
balanced air flow; 9, die clamping assembly**

operating portions of a single screw extruder, as shown in Figure 5-1 and Photo 5-1.

The controlling geometric parameters that characterize the process are the screw diameter, D, and the length–diameter ratio, L/D. Extruders are sorted according to diameter, from which the output of production and the power demand are derived—two dominating operating parameters which depend mainly on the rotating speed of the screw. The diameters in use fall in the range of 1 to 12 inches, but most are in the range of 2 to 6 inches. The ratio L/D is about 24, but has been shifted to larger values (longer screws), of L/D = 36 to 42, in order to improve melt homogeneity.

The extruder may also serve as a pump for viscous liquids, without use of the die. It can also act as a reactor for polymerization or decomposition. In other cases, mostly during the last stages of polymerization or in the compounding process, the extruder is fed with a molten polymer that is compressed and driven out through a special die that generates many strands which are eventually cut into pellets (the pelletizer). However, we will mainly discuss a typical extruder (so-called plasticating) that is fed by solid pellets (or sometimes by a powder) and converts them through internal heating (shearing) as well as external heating (mainly electrical elements) to a melt amenable to flow and shaping in the die.

Therefore, the plasticating extruder has three zones: (1) the feed zone, wherein the conveying of the solid particle is carried out (it is usually cooled in order to eliminate sticking); (2) the compression zone, in which the principal melting takes place; and (3) the metering zone, in which the liquid melt is pumped all the way up to the die. The material itself moves through a helical channel between the screw and barrel. In order to fill up the last portion with molten polymer, the whole system is designed so that the moving volume is gradually decreased from the feed to the exit. In the feed zone (wherein solid particles of a high spatial volume, including air, predominate) there is need for a larger volume. In the compression zone a gradual decrease of volume occurs, by diminishing the channel depth, h (as in Figure 5-2) or by diminishing screw pitch, t. The ratio of the channel volume of the feed zone to that of the metering zone is called the compression ratio, which lies normally in the range of 2–6 (often around 3).

The metering zone is characterized by a constant geometry — comprised of 1/5 to 1/2 of the screw length. In some cases, the whole screw acts as a compression zone (mainly during the processing of rigid PVC). There are various recommended types for different polymers. See Figure 5-2. (The reasons for this variety will be discussed later.)

The advantages of a screw over a ram are obvious and significant, i.e., a continuous process versus a cyclic one, at the same time utilizing the screw for mixing and heating by shearing forces between the screw walls and the barrel. This type of heating is much more efficient than that provided by electrical elements attached to the surface of a barrel which requires conducting heat through a viscous body with a low heat conductivity coefficient. A schematic description of a channel of volume πDbh is given in Figure 5-3.

The rotation speed of the screw is considered the major operating variable, which is capable of continuous alteration. The most common range is 100–

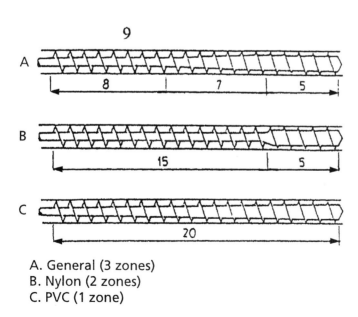

A. General (3 zones)
B. Nylon (2 zones)
C. PVC (1 zone)

FIGURE 5-2 Various types of screws (Note: numbers refer to flights at different zones, respectively.)

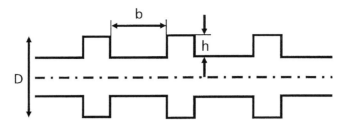

FIGURE 5-3 A scheme of the channel (captions as in Figure 5-1)

200 rpm, but there are special extruders that run at high speeds of 400–500 rpm for specific conditions (adiabatic). It is important to note that shear heating (dissipation) can be expressed by the product of viscosity and the rate of shear squared ($\tau\dot{\gamma} = \eta\dot{\gamma}^2$) with the dimensions of energy per unit volume. Thus, the viscosity and speed (or shear rate) are significant. High speed is reached via transmission between a high-speed motor (up to 1800 rpm) and the screw. The power of the motor in a large machine may reach several hundred kW.

Increasing speed augments the pressure at the end of the screw, essential for the flow of the melt through a die. The pressure may exceed 300 atmospheres.

The pressure is affected by the material properties (viscosity) and the geometry. It can be increased (and controlled) by using a special valve. In many cases a screen is placed prior to the die (eliminating passage of solid polymer or external materials) which also contributes to elevation of pressure.

Contemporary extruders frequently have a venting zone composed of suitable outlets stationed at about 2/3 of the screw length, after which the melt is again pressurized in its advancement towards the die. It is also customary to cool the interior of the screw thus avoiding material sticking to the surface of the screw. (Once it sticks to the screw it will not advance to the exit.)

In some cases (like rigid PVC) there is another advantage to this type of cooling—prevention of degradation. In principle, various screws are used at present for "problematic" polymers like rigid PVC, Nylon and some others. (See Figure 5-2.) In the case of PVC a deep channel is designed, in order to decrease the shear and the resultant heating. In the case of crystalline Nylon, a sudden compression is provided (a minimal compression zone) due to the sharp drop in viscosity during melting. Other modifications exist, like the use of a multiple-flight screw or fins at the end of the screw (or a special torpedo-like structure to improve melt homogeneity).

The average output of a 2-inch diameter extruder is around 50 kg per hour (at a power of 15 kW) but in larger extruders outputs of more than a hundred-fold are reached. The machine efficiency is measured as units of output per kilowatt of the motor, ranging from 3.4 to 8.7 kg/(hr-kW).

Temperature is controlled in several (5 to 6) zones along the barrel, and at the die (separately). Let's now determine the relationship between operating factors (rotation speed, material properties, geometry) and the output, pressure and power requirements. Though the extrusion process starts with solid conveying and melting, we will initially touch upon the process of melt flow, which is most amenable to a quantitative analysis (and therefore has been theoretically treated earlier). In most cases, this comprises the controlling stage of the whole process. Flow of liquid melt occurs in essence in the metering zone, though solid particles may sometimes also be found there, indicating an improper procedure. In the primary mathematical analysis, an approximation of a Newtonian liquid is postulated, which is really inappropriate for polymer melts (although shear rates at the screw are not particularly high). Therefore, when further accuracy is desired, models based on various rheological constitutive functions are presented, as well as the use of numerical solutions. In general, several types of flow of melt in the extruder are defined here.

Drag Flow, q_d

This is the typical flow when a liquid exists in the helical channels between the screw base and the surface of the cylindrical barrel, as a result of relative motion and friction between liquid and metal. As long as nonslipping is assumed, the liquid follows the motion of the screw and a velocity gradient between moving screw and stationary barrel is obtained. However, mathematically it is simpler to assume a rotating barrel and stationary screw. The inclination angle and the friction coefficient determine the amount of material that is dragged by the screw and moves with it. In the absence of friction, the screw would have rotated freely in the liquid without advancing it (which may actually happen with nonviscous liquids). On the other hand, sticking of the melt to the screw will lead to sliding at the surface of the barrel—again not pushing forward.

Pressure Flow, q_p

This type of flow does not occur when the die is missing. In that case, it is considered an open unhampered extruder, serving solely as a pump with only drag flow. However, in common extrusion practice some shaping is always desired, so that a die is used to contract and limit the exit, therefore, building up pressure that reaches a maximum normally at the die entrance. (It may also occur earlier at some point, depending on the geometrical structure of the screw.) This pressure flow is defined as output that may flow back (due to an inverse gradient) the moment the rotation of the screw stops. Thus, the pressure flow makes a negative contribution to the total output of the screw and is termed $(-q_p)$.

Leakage Flow, q_ℓ

There exists a gap (albeit very small) between the flights of the screw and the interior of the barrel allowing some leakage. Therefore, another backflow (in addition to pressure flow) may occur, due to inverse pressure gradient. In practice this contribution is rather minor, as by good engineering design the gap may be made very small, so that q_ℓ can be neglected. However, through use, the gap increases due to aging and wearing of the metal, in which case the screw should be repaired or replaced.

The total output, expressed as volume per unit of time, is obtained by summing the three flows:

$$Q = q_d - q_p - q_\ell \qquad (5\text{-}1)$$

There are two extreme states: First, $q_p = 0$; the extruder is open, and there are no die or pressure drops. In this case $Q = q_d$. Second, $Q = 0$; the extruder is closed (and may be blocked). In this case $q_p = q_d$. This enables calculation of the ultimate pressure, as will be shown later. The ratio $a = q_p/q_d$, possesses an operational significance, and its boundaries are $0 < a < 1$.

Calculation of the flow of a Newtonian liquid in a confining geometry may be mathematically carried out quite precisely; for simplicity assuming flow between parallel infinite plates with a small gap, disregarding the thickness of the walls of the channel. Such an assumption is reasonable for shallow channels, when $(h/b) < 10$ which is generally met in practice.

The final expression follows for the total flow, consisting of drag and pressure components, through the helical geometry. The geometry is illustrated in Figures 5-1 and 5-3.

$$Q = \frac{Ubh}{2} - \frac{h^3 b \Delta P}{12 \mu L'} = q_d - q_p \qquad (5\text{-}2)$$

where
$U \ = \pi DN \cos \phi;$
$L' \ = L/\sin \phi.$
$U \ =$ peripheral velocity in the flow direction (dictated by angle ϕ)
$D \ =$ exterior diameter of the screw (the interior diameter of the barrel)
$b \ =$ width of the channel (fixed in the metering zone)
$h \ =$ depth of the channel (fixed in the metering zone)
$\phi \ =$ angle of inclination of the screw flights
$L' \ =$ length of the helical path of the melt in the metering zone
$L \ =$ axial length of the metering zone
$\Delta P =$ pressure gradient in the metering zone
$\mu \ =$ viscosity (Newtonian in an initial approximation) of the polymeric melt
$N \ =$ rotation speed of the screw

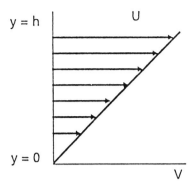

FIGURE 5-4 Velocity profile of drag flow

The angle ϕ plays a significant role in determining the velocity component in the flow direction, leading to a maximum at $\phi = 0$, but in this case drag flow disappears. An optimization calculation indicates that ϕ should be 30°, but due to structural restrictions, most screws are designed with $\phi = 17$–20°.

The velocity component in a direction perpendicular to the flow $V' = \pi DN \sin \phi$, acts only in mixing the melt, increasing with ϕ. Because the product Ubh expresses the moving volume per unit time in the channel, the ultimate yield (in an open discharge) is only 0.5 ($\Delta P = 0$).

$$\frac{Q}{Ubh} = \frac{1}{2} - \frac{h^2}{12\mu U} \frac{\Delta P}{L'} \tag{5-3}$$

This is visualized by drawing a profile of drag velocity in the channel, between y = 0 and y = h, (V = U, maximum) as in Figure 5-4. The triangle represents half of the square Uh (for unit width, b = 1).

In contrast to the linear profile of drag flow, pressure flow shows a parabolic profile, in the opposite direction, its maximum velocity being in the center, as in Figure 5-5.

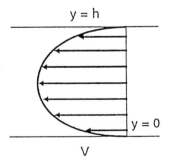

FIGURE 5-5 Velocity profile of pressure flow

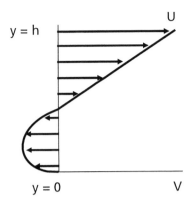

FIGURE 5-6 Velocity profile for
a = 1

$$V_p = \frac{1}{2\mu} \frac{\Delta P}{L'} (y^2 - hy) \tag{5-4}$$

The combination of both velocities produces a negative component in the lower portion, as in Figure 5-6. When Q = 0 (a = 1) both areas are identical.

It is possible to show that for up to a = 1/3, the velocity is always positive (Figure 5-7). In the latter case, Q = 2/3 q_d, so that under these optimum conditions in extrusion through a die, two thirds of the maximum output may be reached.

The ratio, a, depends (among other things) on the operating pressure required for moving the melt through the die. This is expressed in a general way (again for a Newtonian fluid):

$$Q_2 = K\Delta P_2/\mu_2 \tag{5-5}$$

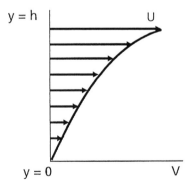

FIGURE 5-7 Velocity profile for
a = 1/3

where

Q_2, ΔP_2 and μ_2 represent the output, pressure loss and viscosity in the die, respectively.

$Q = Q_2$

and in most cases $\Delta P = \Delta P_2$.

K depends on the geometry of the die. For a round die of diameter d and length ℓ,

$$K = \pi d^4/128\, \ell \tag{5-6}$$

An abbreviated expression for the screw equation is:

$$Q = AN - (B/\mu)\Delta P \tag{5-7}$$

where A and B are geometric parameters. The graphical solution is described by a straight-line with a negative slope: the output Q is shown as a function of the pressure gradient ΔP in Figure 5-8.

The maximum value of Q is q_d (as shown before), while the maximum pressure drop ΔP will be obtained when $Q = 0$. For a constant geometry, the parameter A is fixed and parallel lines will be obtained by varying the velocity, so that the output increases directly with N. The slope of the line $B/\mu = (h^3 b \sin \phi)/(12L\mu)$ and so is very sensitive to the depth of the channel (h).

The so-called operating line of the die is a straight-line (as well) but its slope is positive, as shown in Figure 5-9 (that also includes the screw lines). The line starts from the origin, its slope increasing with the diameter of the die, according to the die Equation (5-5). The intersection of the screw and die lines expresses the operation conditions—the output and pressure. It is shown that for a large die, (1), a deep channel should be used (output 1–3 superpasses 1–4), while for a

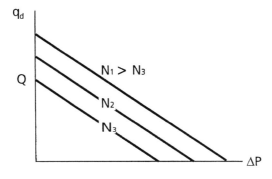

FIGURE 5-8 Operating lines of the screw

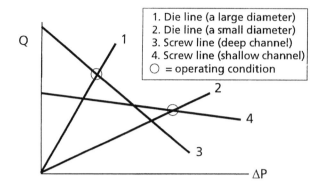

FIGURE 5-9 Operating lines for the screw and the die

small die, (2), a shallow channel is beneficial (2–4). This may also be proved when the thickness h is optimized, as:

$$h, \text{optimal} = (24\ K\ L/\pi D)^{1/3} \tag{5-8}$$

Upon optimizing the angle of inclination, the solution led to $\phi = 30$. Here the optimum angle of $\phi = 30°$ was used (which is not practical) and the screw pitch equals the diameter. A typical expression for parameter K for a round tube was already shown in Equation 5-6. It is, thus, obvious that the thickness h increases with the die diameter, d, but is also affected by other geometrical parameters of both the die and the screw.

Figure 5-9 also shows us that at low pressure, a larger output is obtained for deep channels, while at high pressures, shallow channels provide higher outputs. In general, a shallow channel is preferred due to a decrease in output fluctuations with unpredicted changes in pressure. In addition, the contribution of shearing to heating increases, as the shear rate at the screw may be approximated by U/h. However, the shallow channels are not always in favor, as in the case of rigid PVC (danger of degradation).

By unifying the two equations for screw and die, a general expression is derived.

$$Q = \frac{AN}{1 + (\mu_2/\mu)(B/K)} \tag{5-9}$$

If isothermal conditions are assumed, so that the Newtonian viscosities μ and μ_2 are equal (identical temperature prevails at the barrel and die), a very simple expression is reached (Equation 5-10) as shown in Figure 5-10, where m is a geometric constant.

$$Q = mN \tag{5-10}$$

In this case, the general output depends uniquely on the speed (for a fixed geometry), while the viscosity or even the temperature itself are irrelevant. In

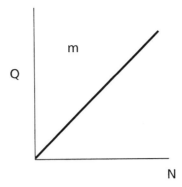

FIGURE 5-10 Dependence of output on velocity at isothermal conditions

the case of $\mu_2 < \mu$ (which will be reached upon additional heating at the die) the output will surpass that obtained at isothermal conditions. In practice it is common to increase the temperature of the barrel towards the die.

The pressure drop depends on geometry, speed and viscosity (temperature)

$$\Delta P = \frac{AN}{(K/\mu_2) + (B/\mu)} \rightarrow A'N\mu \tag{5-11}$$

(for the isothermal case $\mu = \mu_2$). In this case, temperature affects only the pressure drop, whereby ΔP decreases with an increase of temperature, which yields lower viscosities.

Isothermal conditions (either T_1 or T_2) lead to identical outputs, while the pressures are different. It should be understood that the slope of both operating lines increases with an increase in temperature, due to a drop in viscosity (in both cases, in the denominator). This is shown in Figure 5-11.

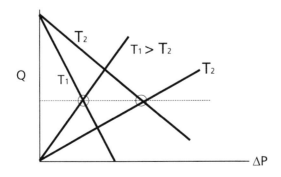

FIGURE 5-11 Operating lines at various temperatures

One of the practical consequences of a basic analysis of extruder operation leads to the possibility of scale-up, namely, the prediction of operating conditions for a large extruder derived with data from a small machine, both having geometrical similarity. Assuming the following relationships prevail

$$D_2/D_1 = h_2/h_1 = b_2/b_1 = L_2/L_1$$

In addition, the rules of similarity also represent both dies, the resultant operation scale-up laws are

(output)	$Q = K_1 D^3 N$	(5-12)
(pressure gradient)	$\Delta P = K_2 \mu N$	(5-13)
(power)	$\Pi = K_3 D^3 \mu N^2$	(5-14)

Π expresses the energy requirement (power) for flow only, $\Pi = Q \Delta P$, which covers only a fraction of the total energy required. It is apparent, at least in theory, that by doubling the screw diameter, both output and power increase eight-fold, while doubling rotational speed the output will also be doubled, but the price will reflect a four-fold increase in power requirement.

Temperature per se directly affects the pressure gradient as well as the required power (both through the viscosity parameter), so that an elevation of temperature will decrease both. Until now we have tacitly assumed (for convenience) that the melt is Newtonian, which appears to be far from reality. Therefore, all the derivations shown so far have served as initial approximations, appearing as the first stage of a theoretical analysis of the extrusion process. (The other assumption of isothermal conditions throughout the process is also impractical.)

It is possible to mathematically solve for non Newtonian flow, provided the exact material functions are known. Even in simple cases, where a power-law relationship exists (see Section 4.1) the resultant expressions are quite complicated, and the drag or pressure flows cannot be separated. However, in this case, the output of an open-discharge extruder is identical to that calculated for a Newtonian liquid. It has also been shown that for scaling-up, the effect of geometry is the same as that for Newtonian liquids, so that for the power-law case under isothermal conditions, the following similarity rules result:

$Q = K_1 N D^3$		(5-15)
$\Delta P = K'_2 N^n$		(5-16)
$\Pi = K'_3 D^3 N^{1+n}$		(5-17)

(Equation 5-15 is equivalent to Equation 5-12.) In Equations 5-16 and 5-17, n is the exponent in the simple rheological equation, the power law, as expressed in Section 4.1:

$$\tau = K\dot{\gamma}^n$$

where n < 1 for shear thinning liquids. Wherever n = 1 (as in a Newtonian liquid) the resultant expressions are similar to the initial ones. Schematic operating lines for non-Newtonian (shear thinning) liquids are shown in Figure 5-12.

In this scheme (Figure 5-12), if we assume that with an increase in pressure gradient (which involves an increase in shear stress), μ will drop (shear-thinning liquid), straight-lines describing the operation of the screw and the die gradually become concave, while the slope increases. In this case, a higher output for the shear-thinning liquid is obtained, similar to flow in pipes. A numerical solution for a power-law liquid flowing in a screw, can be described by the expression

$$Q = \left(\frac{4 + n}{10} \right) bhU - \left(\frac{1}{1 + 2n} \right) \frac{bh^3 \Delta P}{4\mu L'} \tag{5-18}$$

where $\mu = K\dot{\gamma}^{n-1}$ and the approximate shear rate in the screw is expressed as $\dot{\gamma} = U/h$. The latter appears to be of order of magnitude 10 to 10^3 sec^{-1}. Equation 5-18 may be compared to Equation 5-2 where the various geometrical parameters are identical. A two-dimensional analysis, that takes into account cross-flow, provides more precise results, but calls for more computer time.

A generalized analysis that combines the effects of shear, temperature and pressure on the rheological parameters, may lead to an expression of a so-called effective viscosity, $\bar{\mu}$, for a power-law liquid in a limited operational range.

$$\bar{\mu} = \mu_0 \left(\frac{\dot{\gamma}}{\dot{\gamma}_0} \right)^{n-1} e^{-b(T - T_0) + \alpha P} \tag{5-19}$$

where o represents a reference state.

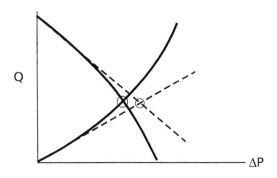

FIGURE 5-12 Operating curves for
non-Newtonian liquids
- - - - - Newtonian flow (straight lines)
———Shear thinning flow (curving)

In contrast to the effect of shear stress or rate, the hydrostatic pressure enhances the viscosity as it decreases the so-called free volume. This case is mostly relevant for injection molding wherein extremely high pressures prevail. In this case a numerical solution is carried out by computer, describing the fields of temperature and pressure for each point. Graphical or numerical solutions may, at best, provide an approximation to real conditions, as the material data are not well known, the parameters apply in a limited range and various assumptions are undertaken for the sake of simplicity.

A complete solution of the extrusion process must include the initial steps, namely: conveying of solids and their melting, the latter representing a tedious target for a theoretical analysis that will eventually fit practice.

Experimental methods for simulating the melting process have also been developed, through freezing and ad hoc analysis. The melt zone initiates when a thin layer of melt appears near the hot wall, eventually followed by a melt pool in contact with the solid bed. The moment the pool fills the channel, the process of melting is completed. The efficiency of melting depends on operating conditions, as well as on heat transfer properties of the system. Until now, we have frequently assumed the existence of isothermal conditions in order to obtain simple solutions, but in reality a significant temperature profile appears during the various heating stages as well as due to local heating by shearing. Sometimes adiabatic extrusion prevails, without any exchange of heat with the environment, (quite important in the case of sensitive materials). During continuous processing, a steady output and a fixed temperature at the outlet of the die are desired, which necessitate a control mechanism. Within the die itself, shear rates exceed those existing in the screw, which may affect the rheological properties (even if the power law is used, the exponent n varies for a wide change in shear conditions). In addition, corrections to the calculation of shear-stress should be applied, the so-called "entrance effects," which become significant with undeveloped flow conditions, as the dies are usually too short. Such a phenomenon is apparent with an increase in molecular weight, due to the contribution of elasticity to the deformation. The viscoelastic behavior produces normal stresses, which show up as melt swell at the outlet. In order to control the desired dimensions of the final product, this has to be taken into consideration in the pulling apparatus. Therefore, there is no identity between the shape of the die and that of non-round articles, due to elastic deformations where (at the outlet) the corners contract more while the center expands.

Even more severe is the appearance of "melt fracture," which occurs during processing at high speeds (above critical shear conditions). It shows up as surface roughness, warpage and eventually as waviness and fracture. The sensitivity varies for different polymers (appearing frequently in linear polyethylene), but rises with molecular weight. This problem may be tackled partly by modifying the die structure (a gradual conical entrance), by varying temperature (a higher one is advantageous) or by changing the material (lowering the molecular weight or utilizing specific additives). However, an upper limit for speed and output is apparent.

The technological development of the extrusion process has led to many practical modifications such as the twin screw extruder which is essential on feeding powders or compounding. The intermeshing twin screw extruder is superior, as it elevates the efficiency of mixing and pumping (acting as positive displacement) and leads to specific uses—processing of polymers that are sensitive to heat, processing of fiber reinforced thermoplastics, and enables controlled processing of thermosetting materials. (In any case, switching from a single extruder to a two-screw machine should always be undertaken after an appropriate feasibility study.)

In addition, processing via coextrusion is becoming popular, wherein two (or more) extruders convey layers of different materials to the same die. The latter method is crucial in the manufacture of multilayered films for food packaging, and for the fabrication of foamed articles by utilizing physical or chemical foaming agents. Additional developments have been made in the processing of extremely viscous polymers of ultrahigh molecular weight using pulsating plungers at very high pressures (which also relates to cross-linked thermoplastics).

Quite revolutionary novelties in extrusion have been offered, wherein the screw is replaced by a series of disks (dispack). The principle in this case is that instead of flow between a static and a moving body, there appears flow between two concurrently moving bodies thus enhancing drag flow, processing capability, and mixing efficiency. The disks are mounted on a common axis and rotate in pairs inside a confining barrel, through which material enters and exits. The dispack machine is suitable for many other processes besides extrusion through a die, such as mixing and compounding, pumping of viscous fluids, as a polymerization reactor, and for separation (devolatilization).

In closing, significant development in the theoretical analysis of the extrusion process has contributed to a better understanding and design of new equipment, an increase of output in existing machinery, and development of concise control via computers.

5.3 INJECTION MOLDING

The processing of thermoplastics by injection molding comprises a major portion of the industry, representing around 50% of processing machinery. The first injection machine was built in 1872 (Hyatt) for the processing of cellulose nitrate. With time, the process underwent improvements and sophistication, so that it may also be used for the processing of thermosets. The distinct difference between injection and extrusion lies in the fact that injection is a batchwise-cyclical process, where the final shape is obtained in a cooled mold, made of two parts that alternatively open and close.

The major utility is for the production of multi-dimensional objects (not longitudinal) with complicated precise shapes including metallic elements, in various sizes starting from minute ones. The common thickness is 4 to 6 mm. In most cases the product injected has the mass of 300 to 3,000 g, but

PHOTO 5-2 An injection molding machine

even 170 kg may be injected in special equipment. Small machines may produce around 70 kg per hour in a screw of 2 inch diameter; a medium machine produces 350 kg per hour with a 4 inch screw. Figure 5-13 describes the mechanism of the process.

The first part resembles in principle an extrusion process with a motor, feed hopper, and a heated barrel containing a screw that melts and positively drives the material. However, operation of the screw is cyclical. It moves forward as a plunger in order to convey a batch of melt into the mold and then retreats through rotation while mixing material and melting a new batch which is then pushed to the front. Actually, prior to use of the screw in injection, a plunger (ram) pushed pellets into a melting zone, in which an

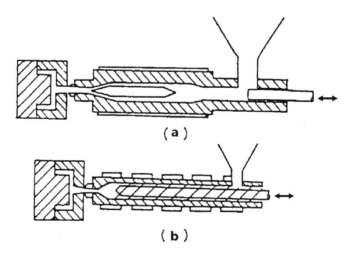

FIGURE 5-13 The layout of an injection molding apparatus:
(a) driven by a ram (with torpedo), (b) driven by a screw

additional torpedo shaped heating element was frequently inserted, mainly serving to separate the melt into thin layers. The plunger also compressed the melt into the mold via a nozzle and a system of conveying channels. Both methods are illustrated in Figure 5-13.

Since the 1950s, the use of a screw has increased, but rams are still mainly found in small machines. A cooled mold is unique in this process, and it is confined by a clamping unit (mechanical or hydraulic) up to 10,000 tons of power, withstanding extremely high pressures. Modern control facilities (with computerized control) enable completely automatic processing while temperatures, pressures and time allocated for each stage are all precisely controlled. The solid pellets are melted both by external heating (mostly electrical) and by shearing forces developed by the screw. A reservoir of melt appears at the front of the screw, which is eventually compressed into the mold, while for a short period, the screw (or ram) holds the pressure and compression during the initial stages of cooling in the mold (thus compensating for contraction by cooling). In order to eliminate back flow, it is customary to use a one-way valve. The compression pressure lies in a range of 600 to 2,000 atmospheres, which calls for a strong hardware system, and extremely high stresses in the mold. The injected dose (metering by weight is preferred to volumetric) depends on the amount of material required for the object in the mold and also the various passages. The system consists of a solid conveying mechanism, melting and pressure flow of the liquid (molten polymer), with two separate zones of heat exchange – the heating zone (melting) and the cooling zone (solidification in the mold). Temperatures reach 160°– 370°C, while cooling is obtained by using water or oil, mostly at 30°–80°C. One has to avoid overcooling, mostly for crystalline polymers that require appropriate conditions for crystallization. In addition, residual stresses should be eliminated during quick cooling, to curtail orientation in the flow direction.

The separation between heating and cooling zones signifies the efficiency of the injection molding process as compared to compression molding of thermoplastics in which stages of heating and cooling alternate in the mold. The ratio between the volume of the melt pool (the hold) and the volume of the injected material determines the residence time. The residence time in the barrel should be minimized in the case of heat sensitive materials especially during processing of thermosets. The advantage of screw over ram lies in an increase in mixing capacity, heating due to shearing, accuracy, mechanical efficiency, and improvement of workability of heat sensitive polymers. It should be understood with ram injection that a large pressure loss occurs in the solid zone, while the efficiency of heating in the melting zone is rather low. One technological solution to this deficiency is based on the use of two plungers, one inclined to the other at a specific angle, wherein the first one melts while the second injects the melt. As with the case of extrusion, the screw consists of feed, compression and metering zones. However, the ratio between the zones and the compression ratio vary for each polymer. The timing control is crucial, where very short cycles of 10–30 seconds are involved. When the mold closes, the screw (or ram) moves forward and conveys a new dose, while holding for an additional compression. During cooling of the mold, the screw is able to retreat to its initial stage before returning to inject a new dose. During this period the product solidifies, reaching such low pressure and temperature that its release is enabled through opening and closing of the mold confined in the clamping unit, without any damage via a rejection mechanism.

It is possible to divide the cycle into three components: first, filling time; second, cooling time; and third, mold opening and closing time. In a specific system, the filling time depends mainly on the flow properties of the melt, that means its viscosity which appears to be both temperature and shear dependent. The filling time may be reduced by increasing the injection pressure. Due to the fact that most polymers behave as shear thinning liquids, it is apparent that the viscosity will drop with a rise in shear rate or stress (resulting from an increase in pressure or a change in geometry), hence the reduction in filling time. Since the process is batchwise, the ratio between dose volume (V) and volumetric flow rate (Q) describes the filling time t_f. An approximate expression for a non-Newtonian liquid is as follows:

$$t_f = V/Q = A\mu_o P^{-\alpha} \tag{5-20}$$

where
A = constant;
P = pressure,
and α is equivalent to $1/n$

In the previous equation, the viscosity μ_o is measured at low shear and is dependent on temperature and hydrostatic pressure. As mentioned before, the hydrostatic pressure elevates the viscosity, and this should be taken into account. It is estimated that a pressure increase of 700 atmospheres, increases the viscosity by approximately 35%. On the other hand, the major effect appears from the shear conditions which reduce the apparent viscosity significantly. The melt viscosities that prevail here are in the range of 10^3–10^4 poise. α is related to the power law ($\alpha = 1/n$), and its value lies in the range of 4 to 5 (as compared to 1 for a Newtonian liquid). This represents the advantage of shear thinning (pseudoplastic) liquids. An effective viscosity may also be used in the following expression:

$$t_f = A\mu_e/P \tag{5-21}$$

The temperature dependence of viscosity may be expressed (in a narrow range), as:

$$\mu_e = \mu_e^o \, e^{-b(T-T_o)} \tag{5-22}$$

where o represents reference conditions (temperature).

If accuracy is needed, the WLF equation should be used, or alternatively, an Arrhenius expression, suitable for the temperature range.

Optimal conditions depend on the sensitivity of the effective viscosity to both temperature and shear conditions. Exact solutions necessitate the use of a rheological equation that fits the material and conditions, as well as the geometry of the flow field, that may frequently be quite complicated. The flow itself takes place in conduits and various outlets (mostly narrow ones) and eventually in the mold itself up to solidification of the melt. Isothermal conditions are not realistic in the mold, while the flow process is hampered by quenching in the cold mold. Unsteady flow may also prevail, as a function of time. With an increase in complexity, numerical solutions are offered to provide the temperature and pressure field at every point, taking into account the freezing after flow. In principle, it is essential to analyze the operating factors and the contribution of the material in enhancing flow. However, flow time consists of only a small fraction of the cycle, while the majority of time is wasted on the cooling process (around 75% of the cycle). Here again, an analytical solution is available for a relatively simple geometry. For example, the cooling time for a thin plate in an infinite plane may be solved, where both walls are at a constant temperature (T_s the wall temperature), utilizing the equation for unsteady state uniaxial heat conduction. The cooling time depends on the boundary conditions, thickness (h), and the thermal parameter, α, called the heat diffusivity. For this simple case, the following expression for the cooling time, t_c, is obtained:

$$t_c = -\frac{h^2}{4.24\alpha} [\log \bar{\theta} - \log 0.81] \qquad (5\text{-}23)$$

where

$$\alpha = \frac{k}{\rho C_p} \quad \text{and} \quad \bar{\theta} = \frac{\bar{T}_f - T_s}{T_i - T_s}$$

k = heat conduction coefficient;
ρ = density;
C_p = specific heat;
T_i = entrance temperature of the melt into the mold;
$\bar{T}_f$ = average temperature of the polymer when mold is opened;
T_s = wall temperature (cooling medium); and
θ = nondimensional temperature.

As mentioned, α is quite small for polymers (due to a small k) and about 10 seconds are required for cooling a thickness of 1 mm, while the time varies inversely as the square of the thickness. One may assume that $T_f \leq T_d$, where T_d represents the heat distortion temperature, meaning the minimal conditions for mold opening without any damage or fracture apply. The cooling time is apparently very sensitive to the thickness of the product and therefore it is not feasible to inject thick objects. On the other hand, it is not customary to supercool the mold, in order to avoid frozen orientation and internal stresses or sudden contractions, and provide free flow in the mold and appropriate crystallization conditions (in the case of injecting a semi-crystalline polymer). In polymers that crystallize quickly, quenching is beneficial as it yields small crystallites which lead to superior mechanical strength and optical properties.

Effective cooling is always desirable, through good design of the cooling channels. In practice, today, numerical solutions provide the temperature field in the mold zone, and computer programs control the cycle time. In conclusion:

$$t_s = t_f + t_c + t_n \qquad (5\text{-}24)$$

where
t_s = cycle time
t_f = filling time
t_c = cooling time
t_n = time for opening and closing of the mold.

The pressure history in the mold during a cycle may be illustrated in Figure 5-14.

Passages leading to the mold (through which the melt flows) consist of a nozzle and a sprue that enables smooth flow but relatively quick solidification

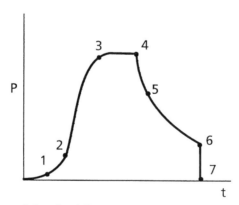

0-1   dead time
1-2   filling of mold
2-3-4 melt compression in mold
4-5   release of back flow (may be eliminated
       by using a unidirectional valve)
5-6   solidification
6-7   opening of the mold

FIGURE 5-14 History of pressure in the mold cycle

PHOTO 5-3 Some products formed by injection molding

thus restricting back flow. From the nozzle, the melt passes through runners that lead to various parts of the mold in a multicomponent system (useful for small items). Premature freezing should be avoided in this case (using deep channels, for example), and a novel solution comprised "hot runners" so that the melt is left in the runners for the next dose and does not solidify (also, this decreases waste). The material proceeds from the runners through a gate into the interior of the mold which provides the final shape. The structure of the gate is important so as to enable quick quenching and a simple and smooth separation between product and superficial material in the various conduits which (for thermoplastics) eventually may be ground and gradually reintroduced into the injection process. The runners should be designed for even flow throughout the mold, especially when a multicomponent mold is designed (mainly for small items). If one considers the large cyclical stresses (temperature and pressure) that occur in the mold and the cooling and flow systems, it is well understood that a successful mold (taking into account changes in dimensions during cooling and solidification, including crystallization) is considered to be a complicated and expensive work of art. One has to remember that contraction in the mold is larger in the case of crystalline polymers (10%–20%) and smaller for amorphous ones (up to 10%). Therefore, the mold represents the most complicated and expensive component in the system, and it will not be feasible to make a new mold unless a production run of tens to hundreds of thousands of objects is planned.

Making a mold is expensive and it should be checked for fit by experimental injection runs, until the exact dimensions are attained. Currently computer aided design and production (CAD/CAM) are very helpful, but the expertise of the mold maker should never be underestimated.

With experience in the injection process, it is seen that each material has a recommended range of workability, as defined by the extreme values of pressure and temperature in injection. (See Figure 5-15.)

Too low a pressure causes an inability to fill the mold or a long filling time. Excessive pressure leads to overpacking the mold, flashing, and severe internal stresses. Excessive temperature leads to thermal degradation of the product, flashing, and at the same time elongates the cooling time. Too low a temperature leads to stream marks, long cycle times or short objects (incomplete filling) in the mold.

An optimal region is recommended for the injection of each polymer. The melt viscosity is sensitive to temperature (energy of activation) and shear conditions, whose rates appear in the range of 10^3–10^4 sec^{-1} (about ten times those in extrusion) in runners of the mold. However, each polymer exhibits a different sensitivity, so that it is frequently useless to heat the system rather than reducing the viscosity through shear. Therefore, it is recommended to increase melt pressure instead of temperature, which may also lead to shortening of the cooling time. (Needless to say, a polymer grade of relatively low melt viscosity is preferred for injection.)

The innovations in injection molding are numerous, and developments are still in progress. Compared to the original process, breakthroughs started with shifting to the use of a screw, larger machines and, particularly, automatic control. Injection of thermosetting polymers has also been introduced, with special design and caution, like a shorter screw and lower compression ratio, strict control of the temperature of the heating region (wherein major heating exists in the mold) and shortening the duration of heating. While cooling the mold is eliminated (actually, taking a hot product from a heated mold), the cooling time is reduced, which represents the major component of the cycle time of injection, in the case of thermoplastics. At the same time, new equipment of injection for reinforced or recycled polymers has been developed.

A special case is the processing of structural foam, including reaction injec-

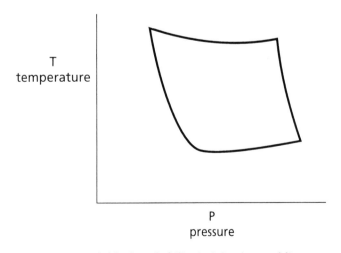

FIGURE 5-15 Field of workability in injection molding

tion molding (RIM), wherein the polymer (mostly polyurethane) is injected while reacting and foaming in the mold. In structural foams objects that are composed of a solid skin covering a foamed core are obtained, when injecting a polymer with a foaming agent to densities in the range of 0.5–0.65 g/cc. The structure obtained yields an extremely high ratio of rigidity per unit mass. It should be mentioned that just by doubling the thickness, the deformation in flexure drops by eight-fold while the mass drops by two-fold. There exists a large collection of equipment at low and high pressures (usually with a melt accumulator with a plunger that causes injection), with multiple entrances and multiple-component systems wherein layers of different polymers are injected in various stages (which actually is coinjection similar to coextrusion), when two injection units are used. The skin is built up by direct cooling near the walls. Common polymers are HDPE, PP and HIPS (high impact polystyrene) with a minimal thickness of 4.5 mm which produce a foamed structure resembling wood and, therefore, also replacing it. In the RIM method, both full and foamed bodies are manufactured. In this case, high pressures or temperatures are not needed, which leads to a major saving in the mold price, with aluminum or epoxy used in addition to steel (the pressure is about 4 to 8 atmospheres). The monomers feed undergoes rapid polymerization and cross-linking within about 30 seconds out of 1–2 minutes of cycle time, reaching an output of 5 kg/sec. Accessories are used for metering, conveying and mixing the various components that interact, in order to produce polyurethane in the mold. In addition to polyurethane, other polymers may be processed by RIM, like unsaturated polyesters, epoxy and urea, as well as thermoplastics like polyamides. (A rather novel equipment is manufactured by Teledynamic, which processes by pulses at high pressures and may be helpful for poorly workable polymers.)

There are various simulators for polymer workability. A popular one consists of a spiral mold that enables testing the various conditions (melt temperature, injection pressure and temperature of the mold walls) that produce an effect on flow time before freezing. The measured parameter is the length, where a good correlation is found between length and melt viscosity at the high shear rates prevailing during injection. Upon increase of the temperature in the melt or of product in the mold, or increase of the pressure, the helical length increases as well. However, the resultant values and sensitivities vary for each polymer, the behavior of which depends on structure, chainlength and distribution.

5.4 BLOW AND ROTATIONAL MOLDING, THERMAL FORMING AND CALENDERING

Blow Molding

Whenever hollow objects are desired, like bottles and containers, blow molding (a modification of injection molding) is used. In this process, an extruder usually continuously provides a melt cylinder (parison), in such a manner that its production rate fits exactly the time for catching and cutting the parison and closing the mold. In the mold, the parison is expanded by air pressure and takes the shape of the mold walls with a desired thickness, then cools and ejects from the mold. It is possible to increase efficiency, by using a series of rotating molds, all fed by a single extruder. The appropriate synchronization between the continuous and periodical processes is crucial. In a compact unit, the extruder is a part of the machine and a modified screw combines a piston-like motion with a rotational motion in a cyclical process, (which is useful mainly for the production of large objects, up to 5,000 liter). Another option consists of a melt accumulator with a ram or plunger that creates the parison. In another process, (injection blow molding) the blow molding consists of a direct stage of injection on a steel bar which is transferred to blowing. In this system, better automation is reached, in addition to a saving in material (which is wasted as surplus in ordinary extrusion) where the parison is provided with extra material which is twisted to seal prior to blowing. These methods are suited mainly for small hollow objects with a volume of up to half a liter. The molds do not withstand high pressures and the blowing pressure lies in the region of 5–7 atmospheres. The cycle time is about 15–60 seconds (depending on thickness and cooling efficiency). The temperatures are lower than in injection, as there is no requirement of low viscosity for flow, rather than a plasticity that enables deformation. On the contrary, the parison must be viscous enough to retain its stability as a free vertical structure prior to blowing. This is related to the melt strength which calls for higher molecular weights than in injection, particularly when the parison is produced by extrusion. The flow is in essence viscoelastic where expansion of the free jet must be taken into account, which eventually also determines the thickness of the product. Blowing in the mold represents flow induced by normal stresses (with no shear stresses). This is characterized by the elongational viscosity which differs from the conventional viscosity. The rheological analysis of flow, combined with stretching, is rather complicated. There exists a fabrication process for bodies of varying thicknesses at different zones (with local strengthening and saving material). Here the thickness of the parison is varied either through altering the flow rate, that effects blowing at the exit or directly in a mechanical way.

The blow molding process has undergone many developments since its early appearance in the 1920s with the processing of cellulose products by methods borrowed from the rubber industry. One of the successful developments consists of simultaneous blowing and stretching in order to attain a two-dimensional orientation (stretch blow molding) that has led to the production of bottles made from thermoplastic polyester (PET) which exhibits extraordinary properties like excellent mechanical strength and sealing of vapors. PP and PVC may also be fabricated by this method, and multilayered objects are also made. Hollow objects are thus manufactured from polyethylene (mainly HDPE), PVC, PP and PET. In essence most thermosets can be blow molded either through extrusion or through injection, as well by a combination of both methods. It is possible to

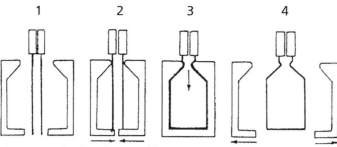

1) Entrance of parison (open mold)
2) Mold closes
3) Expansion by air pressure
4) Open mold (product ejected)

FIGURE 5-16 The principle of blow molding

manufacture bottles and containers directly on-line, and accomplish filling by specific liquids. A scheme of the blow molding process is shown in Figure 5-16.

Film blowing (Figure 5-17) represents one of the most useful extrusion processes. It is based on a free sleeve that is made by blowing with an air bubble, at an interior pressure of 0.05–0.2 atmospheres. The blow ratio ranges from 1.5 to 4, and this parameter determines mechanical and optical properties. In addition, it is possible to gain biaxial orientation which increases the strength or uses elastic memory for shrink packages (by heating). In this case, the speed of the stretching roll enhances orientation in the machine direction, while an increase of air pressure enhances orientation in the axial direction. The cooling system and its location are essential for determining product quality, while the speed of

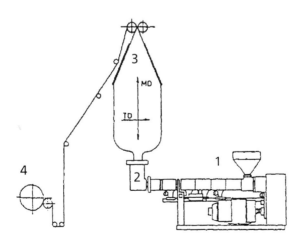

FIGURE 5-17 The process of extrusion film blowing
1) Extruder 2) Die 3) Blown sleeve 4) Rolled films
MD = machine direction; TD = transverse direction

stretching determines the final thickness of the film. Very thin films in the range of 10–50 microns can be manufactured, where the typical range of films is limited to 0.25 mm. The most common method uses a special die (at an angle of 90°), providing an annulus of melt that is blown by an air bubble into a vertical sleeve. In large machines, very wide films are produced (around 20 meters) carefully retaining homogeneity with a rotating device at the exit (a rotational cooling ring or rotational die). There exists also a horizontal blowing system for the production of films by means of casting a thin layer on a chilled metallic wall. Today, it is customary to produce multilayered films (mainly for food packaging) by means of coextrusion using a number of polymers (including recycled material).

Rotational Molding

Like blow molding, here again hollow objects are manufactured, but the method of rotational molding varies in several respects. First, the feed is mainly a powder that is weighed directly into the mold. Second, the mold itself is rotated (along two perpendicular axes) and heated until the polymer melts and is pushed to the walls for gaining the shape. During the last stage the mold is cooled and the product ejected. The powder initially undergoes a process of sintering followed by melting.

The advantage of rotational molding stems from the possibility of processing very large and thick objects (up to 12 mm), including foamed structures and multilayered ones. The most common polymer processed by the rotational molding method is polyethylene, but other polymers are also used, including plastisols of PVC, from which toys were already manufactured in the 1940s. The whole system is heated in an oven (also possible with an open fire) at a temperature range from 200°C to 400°C and a speed of 5–20 rpm (sometimes higher speeds are used). The polyethylene powder is obtained by mechanical grinding into a size region of 300–1200 microns, with a common particle size of 500 microns. The cycle is relatively long and therefore is restricted to fabricating very large objects (containers up to 10 m^3) or hundreds of items in a single run. In this method, very little material is wasted, and orientation is also eliminated. Due to the use of low pressures, cheap and thin molds may be employed— which aid heat transfer. The method is applied also to the processing of thermosetting and reinforced materials, and to a host of thermoplastics like polystyrene, Nylon, polycarbonate, ABS and acetal. In any case, the range of molecular weights is low compared to those met in extrusion or blow molding. A similar processing method involves coating by a polymer powder melted in a fluidized bed.

Thermoforming and Vacuum Forming

Forming represents a relatively low cost method, utilizing male or female molds, to which a heated extruded polymeric slab is attached, so that the desired shape is obtained mainly through vacuum or positive pressure. (See Figure 5-18.)

During the final stage the product is cooled until it solidifies. Temperatures are moderate as the material mainly undergoes viscoelastic deformation rather than pure flow. The temperature range is slightly above T_g or T_m (120°C–230°C). By this method one obtains precise and large objects of relatively thin cross section, as refrigerator doors, baths and structures up to 4×10 meters. Another broad utility is in packaging and production of small items. The process is batchwise but rather quick (several seconds), and automatically controlled. Common polymers that are fabricated by this method are HIPS, ABS, PVC, acrylics as well as many others.

Forming may also be used for direct packaging of objects, continuous extrusion of an infinite sheet which is formed on-line, or a combination of injection molding and thermal forming. The process itself is composed of three major steps–heating, forming and cooling. Whereas heating of plastic sheet by conduction is not too efficient, various methods of heat transfer have been developed, the most prominent one being based on infrared radiation. In any method, a careful control of the desired temperature should be employed in order to eliminate material degradation, so that hot air or heated plates are preferred. In forming, both male and female molds are common, utilizing a plunger and thin low-cost molds. The pressure is rather low (air pressure reaches 10 atmospheres), but vacuum by suction through holes in the mold is preferred. With deep objects, the thickness may drop rapidly as it gradually decreases along the plunger stroke. The thickness, however, can be controlled and zones of different thickness can be reached through sophisticated methods. Slabs as thick as 2 mm are provided as long cylinders for continuous feeding. Thermal history affects the appearance of residual stresses and frozen

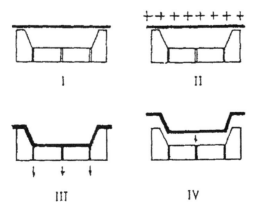

FIGURE 5-18 Scheme of vacuum forming operation
I) Attachment II) Heating III) Performing vacuum and cooling IV) Ejection of final product.

orientation, taking into account also the previous processing step in extrusion. Upon release of the stresses, undesired contraction of the product during heating may be eliminated. Usually there appears another step of trimming and finishing, so as to remove edges and excess material.

Calendering

Calendering is a continuous fabrication method like extrusion. It has been used for making films and coatings for a long period, starting in the rubber industry and penetrating into the processing of plasticized PVC and the coating of textiles and paper. The outline of this method is shown in Figure 5-19.

The core of the machine is a series of polished and heated rolls (1) between which the material passes after stages of compounding and mixing that follow the extrusion of strands which are eventually converted into sheets of thin and even thickness. The rolls themselves are set in a special pattern (inverted L, Z and others) and usually four, but sometime five, rolls at various speeds and temperatures are encountered. Material that ejects from the last pair of rolls, reaches a final shape and passes into stretching (5) and cooling rolls (4), and eventually to collecting rolls (6). The whole process is running continuously. The thickness of the product is currently controlled by gamma rays. The material is compressed into a gap between each pair of rolls, so that in the whole system of four rolls there are three passes providing feed, metering and forming of the sheet. (The latter dictates the thickness.) The transfer from roll to roll is obtained by a gradient in temperature or velocity. The range of thickness is broad: 0.05–0.75 mm, but in flooring, a thickness of several mm is reached, at an output of 30 to 120 meters per minute (up to 5 tons per hour), at a width of 1.5–2.5 m, although a width of 3.4 m can be reached. The power requirement is less than in extrusion, and this operation mainly fits the processing of quite highly viscous polymers like PVC that are sensitive to thermal history.

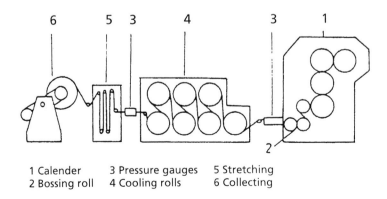

1 Calender   3 Pressure gauges   5 Stretching
2 Bossing roll   4 Cooling rolls   6 Collecting

FIGURE 5-19 Principle of calendering

Control of temperature (150°C–250°C) and even thickness is most important. With increasing temperature, both viscosity and power requirements are reduced, but high temperatures cause thermal degradation and a drop in melt strength and mechanical stability. Shearing between rolls also adds thermal energy, so that conditions are not isothermal. The thickness of a sheet passing through a fixed gap between rolls varies, so that it decreases from the center to the edges as a result of hydrodynamic forces. Therefore, the rolls are constructed, crossed or bent, so that the diameter at the center is slightly larger than at the edges.

The power requirement depends on the geometry of the roll, speed and gap, as well as on polymer viscosity. In spite of an initial high capital investment and a relatively long time to reach a steady-state, calendering offers an apparent advantage for the fabrication of sheets at a high rate with excellent quality. A mathematical analysis of the flow process in a calender has been developed for both Newtonian and non Newtonian liquids, including viscoelastic behavior, whereas, in simple cases an analytical solution is possible. Through approximations a numerical solution by finite elements is derived, providing very accurate information about the field of temperatures, pressures and the effect of velocity on thickness and output. Contrary to extrusion, there appears to be almost no orientation. The main products are sheets and fabrics for upholstery (printing and decorating), conveyor belts, and flooring.

Other Fabrication Methods

Casting   In this method, the material is cast as a liquid into a mold wherein the object solidifies and reaches its final shape. A common process consists of feeding a liquid monomer that polymerizes in this manner, mostly for acrylics, Nylon, polystyrene and polyester. In addition, PVC plastisols (high concentration of plasticizers) that appear as liquids at ambient temperature, harden (gel) after being cast and heated (the result of solubilization of the polymer in the plasticizer). Other cases of casting include thermosets. A modification of casting is plunging a male mold into a liquid in order to obtain various products such as gloves (PVC or rubber) and toys.

Cold Forming   This method resembles that involving metal processing and requires high pressures, but eliminates thermal history and extensive cooling stages. Most commonly used for pressing and pulling is ABS. Another utility is pressing of tablets as a means of feeding thermosets into compressing processes.

Foaming   Production of foams via extrusion or injection, as well as by other processing methods, has already been mentioned. The purpose is to construct a hollow structure of low density, thus improving thermal and mechanical properties, in addition to being cheaper. There is a distinction between chemical foaming, where a special component is added that may be decomposed by heat or that reacts chemically (like in polyurethane), and physical foaming, wherein a gas or volatile liquid is added at the last stage (like in PS). (In chemical foaming, a specific mass of about 0.5 is obtained, compared to up to 0.03 for physical foaming.)

In general, there are five major foaming processes:

1. Mechanical — frothing.
2. Generation of foam by exhausting gas from a chemical reaction.
3. Dissolving and rejecting a soluble component.
4. Introduction of a chemical or physical foaming agent into the polymer melt.
5. Sintering.

(Sometimes a nucleating agent is added, to control the size and distribution of pores.)

In injection, high pressure is advantageous in eliminating premature foaming, which is more difficult to control in extrusion. Foamed polymers include polystyrene and polyurethane, as well as urea-formaldehyde, in addition to most useful polymers (including engineering). In the case of PS, pellets containing a foaming liquid (often pentane) are usually prepared and processed in two steps, an initial step consisting of early foaming the swollen particles. Final foaming is performed at a low pressure of about 5 atmospheres, with a source of heating being steam in most cases. A rigid foam of specific mass in the range of 0.015–0.5 is obtained. In the case of polyurethane foams, one has to distinguish between rigid or flexible foams (details in Chapter 6). Here we deal mainly with casting or spraying of two major components; one of which may include the foaming agent (like Freon). After mixing, a chemical reaction occurs between the polyol (hydroxyl) and isocyanate, evolving CO_2 (in the case of flexible foams), and releasing heat that evaporates the foaming liquid (in the case of rigid foams). The gas is formed during fixing of the structure by cross-linking, so that a stable structure of closed pores (rigid) or open ones (flexible) is reached. Wide sheets may be produced on a conveyer belt, or the material may be injected into a mold. The range of specific mass is as follows: for rigid foam, 0.03–0.3; for flexible foam, 0.015–0.045.

5.5 PROCESSING OF THERMOSETS

It has already been mentioned in previous paragraphs that currently most fabrication machines designed for thermoplastics may also be modified for thermosets. However, a crucial problem is to avoid premature solidification prior to final forming, so that processing must take place during the plastic stage before cross-linking. Care must be taken to eliminate excess material that cannot be reworked. This calls for careful and strict design. In this section some processing methods primarily used for thermosets will be described.

Compression Molding

This is an ancient method already used in the rubber era, but still very useful for many thermosetting polymers, such as phenol, urea and melamine formaldehydes, unsaturated polyesters and epoxy. Sometimes it may also be adopted for the preparation of sheets and laminates (multilayered) from thermoplastic materials, as well as to produce records. The compression molding machine is based on a hydraulic press providing the pressure, and a heated mold made of two parts. For thermoplastics, a cooling mechanism is also applied, which increases the cycle time. The process itself is typically batchwise, and sometimes may take several minutes prior to the end of cross-linking (times range from 15 seconds to 5 minutes). Usually feed consists of preweighed doses of molding powder, but it is preferred to feed ready-made tablets, manufactured primarily by cold pressing. The processing stages are as follows–measuring feed into the mold, closing the mold, heating and fusing the polymer, application of pressure, hold-up for curing, and finally opening the mold and ejection of the product (no cooling required). Common conditions are temperature of 140°C–200°C, pressure of 120–600 atmospheres. Heating is obtained by steam, oil or electricity. Heat transfer is not too efficient and there is danger of overcuring in thin parts and undercuring in thick zones (both undesired for shape and strength). Preheating the feed is currently preferred. Emission of excess polymer and vapors should be allowed for. In addition, one has to take into account shrinking of the material. In order to increase the efficiency of the compression molding process, multicomponent molds and the use of extruders for feed have been introduced— thus leading to full automation. Due to the high pressure that prevails at the mold, bodies that contain inserts are difficult to fabricate.

Comparing compression molding to injection molding, the former comprises a longer cycle, in addition to difficulties in the fabrication process of delicate objects. On the other hand, compression molding is cheaper in cost of machine and molds, and there is less waste of material.

Transfer Molding

In compression molding, high pressures are applied at the mold, frequently hampering dimension stability of the product, mainly when more elements are introduced. In the method of transfer molding, the system combines injection with compression. There is a heating zone in which a dose of feed is plastified and the liquid is eventually injected through a gate into a closed mold, in which the appropriate pressure and temperature for curing prevail. At the end of the process, the mold opens and product is ejected. The transfer zone is located directly above the mold. As the material is conveyed to the mold as a liquid it flows better and this permits complicated and delicate structures, including metallic elements. This method is more efficient and quicker than compression molding.

5.6   PROCESSING OF REINFORCED MATERIALS

In this discussion, specific methods will be dealt with for processing polymers, which have been reinforced mainly by the addition of fibers. One should remember that all thermosets require some sort of reinforcement, but currently many thermoplastics are reinforced, as well, particularly those defined as engineering polymers.

Most reinforcement is accomplished by various types of glass fibers. In many cases, manual labor is still used for processing, but there already exist many sophisticated machines that fabricate products of high quality and reproducibility, mainly by methods of compression but also by injection and extrusion. The simplest method is termed hand lay-up, in which glass fibers (mat or fabric) are wet with the resin (usually unsaturated polyester or epoxy) and hand laid as several layers on top of a core layer of resin which is spread by means of a brush or roller on a simple and cheap mold. Care must be taken to eliminate air bubbles and to densify the material prior to curing at room temperature or in an oven (the latter is preferred). A more advanced method is based on spraying a mixture of resin and chopped glass fibers by means of special spray guns (for boats, for example). More advanced fabrication methods are comprised of pressure molds of various shapes, built as two fitting parts. The feed may consist of a precursor made of glass and resin, or a viscous paste that is able to flow into the mold. Other methods are based on the use of an elastic expanded sack that develops pressure (or vacuum) by which the material is pushed into a hollow mold where it solidifies. Air is easily removed and very smooth surfaces are achieved (as in production of boats) with a high strength-to-weight ratio. The main problems lie in achieving an even spread, the quality of bond between polymer and fiber, the utilization of long fibers, and the elimination of surface defects. During continuous processing a conveyor belt is used, by which the reinforcement is moved (glass fiber or other fiber types, including paper) while the resin and catalyst are sprayed, and the whole system is cured and hardened by movement through an oven. The belt velocity is in the range of 1 to 10 meters per minute. Flat or corrugated sheets may be obtained, (by utilizing shaping elements prior to crosslinking) at a thickness of 0.5–5 mm and width of 1 to 3 meters.

More advanced continuous methods consist of filament winding and pultrusion. In filament winding, fiber is soaked with a liquid resin and continuously winds around a rotating axis (eventually removed after curing) in order to achieve symmetrical structures (round, conical and others). By this method the reinforced structure gains its best advantage, enabling automatic and sophisticated fabrication for specific targets such as pressure tanks and various elements in transportation and aviation. In pultrusion, fiber soaked with resin is transferred through a heated die that provides the desired shape during curing and pulling by special equipment. In both methods, very high concentrations of continuous fibers may be used (up to 80% by weight) for engineering and constructive outlets. In modern methods of compression molding of reinforced polymers, a ready-made paste is used consisting of resin with all the additives (including chopped-fibers, catalysts and thickeners), DMC (dough molding compounds), or SMC (sheet molding compounds). In both methods, excellent mechanical properties are achieved, however, SMC has become more developed, consisting of sheets ready for compression. This method may also use continuous fibers. On the other hand, DMC (sometimes also termed bulk molding compound, BMC) is cheaper and also suitable for complex shapes. It is possible, currently, to overcome the large shrinkage of polyester (about 8%) by using a low profile system incorporating a small portion of thermoplastic polymers.

5.7 WELDING AND BONDING

This discussion addresses to the finishing operations which combine objects. In principle, thermoplastics may be bonded either by welding (like metals) or by gluing. We will consider the most useful welding methods.

Welding by Hot Gas   A gun constructed like a welding hammer is used, through which flows a stream of electrically heated air or nitrogen. The temperature reaches 200°C–320°C. (It is mainly used for PE or PVC.) A welding bar, made of the identical polymer is commonly used to fill in the binding area. One must avoid degradation.

Welding by a Hot Plate   In this method, two parts are bonded through contact with a heated metallic sheet, (120°C–200°C) during a preplanned contact time. It is useful for polyolefins and vinyls, and mainly for the fabrication of packaging bags.

Ultrasonic Welding   A quite efficient method is based on ultrasonic energy, particularly for the bonding of thin films. It is useful for polyester (Mylar) and others. The heating is very rapid.

Welding by Friction   This method exploits the heat of friction when two cylindrical bodies counter-rotate. This method is sometimes used with rigid PVC, Nylon and acrylics.

Welding at High Frequency   Dielectric heating is accomplished with electrodes that are connected to a generator of voltage at high frequency. This method is very efficient but is restricted to polymers that exhibit a high electrical loss factor like PVC and other polar polymers.

Bonding   The use of solvents (mostly with a cement made of a viscous solution of the same polymer) is common for easily soluble amorphous polymers like acrylics, polystyrene, cellulose, and PVC. However, this method does not apply to most crystalline polymers. The proper choice of solvent (or mixture of solvents) also takes into account the desired rate of diffusion. In this matter, the concept of solubility parameters may be exploited when there is a similarity between polymer and solvent. Various glues may also be used (contact glue, epoxy) in order to bind plastics to each other, including crystalline polymers or thermosets.

Another combination method is mechanical. It consists of using screws and pins, or ball and socket joints, wherein the elasticity and contractability are utilized for successful contact. Milling of most polymers resembles, in principle, that of soft metals, but in the case of rigid polymers care must be taken against cracking and intrusion of mechanical stresses. In design, sharp corners should definitely be avoided.

5.8   COMPARISON AMONG VARIOUS PROCESSING METHODS

The processing of plastics is based on three major operations:

1.   Heating to soften (or melt) the material.
2.   Shaping (flow).
3.   Cooling the product and release of heat.

In the case of thermosets, the product may be ejected while hot and, eventually, cooled at ambient temperature.

In this chapter, various processing methods were described and analyzed, their major function being shaping. The dominant continuous methods are extrusion and calendering. The common cyclical methods are injection molding, blow molding, thermal forming, and rotational molding. In general, the shape of the product dictates the choice of processing method—elongated objects are processed by extrusion, hollow objects by blow or rotational molding. In addition, material restrictions and behavior must be taken into consideration (like the sensitivity of PVC to thermal history, or processing difficul-

TABLE 5-1

Comparison among Major Processing Operations

Processing Operation	Minimal Feasible Production	Initial Machine Cost	Initial Mold (Die) Cost	Output	Cycle Time	Melt Temperature	Working Pressure
Injection molding	10^4–10^5 items	high	high	high	small	high	very high
Blow molding	10^3–10^4 items	low	low	high	small	high	low
Rotational molding	10^2–10^3 items	low	low	low	high	high	low
Extrusion	300–3000 meters	high	low	high		high	high
Thermal forming		low	low	high	high	low	low

ties with Teflon). Sometimes it is possible to produce the same object by several processing methods. In this case, different effects on final performance and properties (like orientation) should be considered, as derived from the different flow and cooling conditions of the distinct processing methods.

In methods of extrusion, injection and rotational molding, the feed consists of pellets or powder. In the case of blow molding, an intermediate product (the parison) is made, and prefabricated sheets or films are utilized in thermal forming. (Therefore, these two processes are considered as secondary.) In the major primary processes of extrusion and injection molding all stages (starting with solids and ending as solids), are present; that is, solid conveying, melting, melt compression and flow, and cooling. Usually there appears also a stage of post-fabrication finishing.

When different processes are compared, economics is of vital importance in determining feasibility. Currently, the requirement to reduce energy consumption is most important. Table 5-1 provides data for comparison.

PROBLEMS

1. a. In injection molding it was found that the product as obtained is too short. Suggest three possible solutions. What are the limitations?

   b. How will cycle time be affected by doubling the thickness of the product? By decreasing temperature of cooling water? By increasing pressure? Comment.

c. What is the advantage of injection molding over compression molding?

d. What is the advantage of the screw over the ram in injection molding?

e. What is the advantage of using a "hot mold" in injection molding?

2. Determine the cycle time of an injection-molding process of polystyrene at the following conditions. How can you decrease flow time? Cooling time? Dead time?

Temperature of injected polymer: 230°C.

Temperature of the polymer ejected from the mold: 77°C

Wall temperature of the mold: 65°C

The product is a thin plate, with thickness of 0.07 inch

Heat diffusivity $\alpha = 10^{-3}$ cm^2/sec.

Filling time $= t_f = A\eta_o P^{-\gamma}$ (time in seconds and P in kilo psi).

Low shear viscosity $= 34,000$ poise.

$A = 8$

$\gamma = 4.5$

Pressure: 12.8 kilo psi

Assume dead time of 6 seconds

3. If cooling time represents 75% of the total cycle time in injection molding, determine the change in output if the thickness of the product is doubled.

4. Compare processing by blow molding to injection molding in regard to the following variables: (a) temperature; (b) pressure; (c) molecular-weights (viscosity).

5. What is the advantage of processing by rotational molding as compared to injection molding? What are the limitations?

6. a. What type of objects are made by extrusion or by injection molding?

b. What is the purpose of the torpedo in injection molding?

c. What is the difference in the operation of the screw in the extruder and injection molding?

d. What is the importance of the screw length?

   e. What is the importance of the screw angle in extrusion?

   f. How can the efficiency of cooling the mold be improved in injection molding?

7. A certain extruder operates at 35 rpm. Under these conditions the maximum flow rate is 50 liters per hour and the maximum pressure is 500 atmospheres. When a cylindrical die is attached, of diameter 3 mm and length 30 mm, calculate the actual production rate and pressure. Assume Newtonian melt viscosity of 10^4 poise.

8. An extruder operates at 100 rpm. The internal diameter of the barrel is 2 inches, and total L/D is 16. Assume that the metering zone occupies one quarter of the screw length. The screw angle is 30°, the channel width is 1 inch, and depth 0.1 inch. The temperature in the barrel and die is constant at 330°F.

   Calculate the following:

   a. Maximum capacity of the extruder.

   b. Maximum pressure.

   c. If speed is doubled, while the die is cooled to 280°F, how will the flow rate by changed?

   d. If both barrel and die are heated to 350°F, without changing the screw speed, how will the flow rate and pressure be affected?

Data:

Melt Viscosity (Newtonian)

at T = 280°F, 0.04 lb·sec/inch^2

at T = 330°F, 0.02 lb·sec/inch^2

at T = 350°F, 0.01 lb·sec/inch^2

9. A sheet of plastic is made continuously in a screw extruder. The dimensions of the sheet are: thickness, 5 mm; width, 1.2 meter; and production rate of 25 cm length in 5 minutes. The extruder has the following parameters: diameter, 50 mm; width of channel, 25 mm; depth, 2 mm; angle, 30°; speed of screw rotation, 75 rpm (rounds per minutes).

   a. Is it possible to reach the desired production rate under the prevailing conditions?

   b. If not, what is the minimum screw speed for this operation?

   c. How will the change affect power requirement?

(Hint: Calculate the maximum possible output.)

10. A laboratory extruder of 1.5 inch in diameter is used at a speed of 100 rpm. What will be the ratio of output, pressure and power when an industrial extruder of 4 inch diameter is applied (both in geometrical similarity), working at 150 rpm. Assume no change in melt temperatures.

6

Description of Major Plastics: Structure, Properties and Utilization

*T*HIS CHAPTER compiles the information available in the field of polymers and plastics regarding structure and properties, and indicates significant unique uses. Special developments within each family of polymers will be highlighted. About 40 families of polymers will be briefly described, appearing commercially in over than 13,000 grades. After a detailed description, comparative data on properties, utilities and prices will be provided. It is obvious that the data represent typical average values, as there appear many grades of each polymer differing by molecular weight, distribution, degree of crystallinity and so on, not to mention modifications, blends and copolymers. In addition, the strength, rigidity, and sometimes also toughness of the polymer can be surpassed by appropriate compounding, mainly by reinforcement. The order of presentation of the materials is ranked according to quantitative utilization.

6.1 INFORMATION ABOUT VARIOUS PLASTICS

Forty families of useful plastics will be surveyed, subdivided traditionally into thermoplastics and thermosets. It is roughly possible to sort the polymers according to decreasing order of usefulness and increasing order of raw material price. As the price diminishes the rate of use rises or upon the increase of sales capacity the price of raw material may drop. The first group concerns commodities at the price range (1990) of 40–60 cents per pound and a high utility (above a million tons annually in the U.S.). This group consists of polyethylene (LDPE, LLDPE and HDPE), polypropylene, PVC and polystyrene. Recently, due to enormous growth in its use as packaging for containers and bottles, PET (saturated polyester) may also be considered to belong to the commodities materials.

Intermediates consist of ABS, SAN, PMMA (acrylics) and engineering derivatives of cellulose like CAB (cellulose acetate butyrate), their relative price reaching around 2 (on the basis of PE as unity). Some consider these polymers at the low range of engineering polymers, due to their physical properties. (PP is frequently sorted as such.)

The engineering polymers that have already reached maturity consist of the Nylons (PA), polycarbonate (PC), acetal (POM), polyesters (PBT and PET) and Noryl (PPO). Their relative price is around 3. Including very novel polymers, a prestigious high priced group consists of the advanced engineering polymers (high performance); polysulfone (PSU), polyphenylene-sulfide (PPS), fluoroethylenes (PTFE and its derivatives), polyamide–imide (PAI), polyetherimide (PEI), polyethersulfone (PES), polyether–ether–ketone (PEEK), aromatic polyesters and polyamides, polyarylates and liquid-crystal-polymers (LCP). Here the price ratio reaches 10–50 (sometimes even more) but the performances are outstanding. Each polymer will be described in detail, followed by tables and data regarding properties, uses and prices. From year to year prices vary with surprises and fluctuations, but the relative rate of consumption is the determining factor.

Thermoplastics

1. Low Density Polyethylene (LDPE)   LDPE (specific mass 0.915–0.935) was until recently the most common polymer; however, some part of the market is being transferred to LLDPE (linear low density PE). LDPE was developed by ICI in England in 1933 and its production started in 1939. The chemical structure represents an addition product of the gas ethylene $CH_2=CH_2$ leading to $(CH_2-CH_2)_n$. This is a typical product of the petrochemical industry, where the monomer ethylene (at 99.9% purity) is mainly obtained by cracking of gases at the crude oil refineries. The polymerization method is based on a continuous reaction under high pressure (in the range of 1200–3000 atm) and temperature of 300°C carried out in reactors like stirred-vessel autoclaves (ICI method) or tubular reactors (Hoechst method). Each method has its advantages and disadvantages, while the modern industry switches to huge tubular reactors that can produce over 300,000 tons annually. The governing mechanism is radical, where oxygen or peroxides serve frequently as initiators.

The reaction is very rapid (seconds) and conversion to polymer reaches 10–30% (to avoid excessive heating), followed by separation of a mixture of polymer and monomer, the latter being recycled in the process. The product, PE, is conveyed as a melt into an extruder then emerging as pellets. The range of molecular weight is:

$$\overline{M}_n = 15{,}000 - 40{,}000; \ \overline{M}_w = 10^5 - 10^6$$

so that the distribution is apparently quite broad

$$\overline{M}_w/\overline{M}_n = 5 - 40$$

From the molecular geometry LDPE is structured as a branched polymer, consisting of short and long branches. The short ones (at an average frequency of 20–40 CH_3 groups per 1000 C) are based on side chains of C_2–C_6, which tend to decrease crystallinity to the range of 40–60%, thus determining most mechanical and thermal properties.

$$-CH_2-CH-CH_2-$$
$$|$$
$$CH_2$$
$$|$$
$$CH_2$$
$$|$$
$$CH_2$$
$$|$$
$$CH_3$$

The long branches (average frequency of 2 to 4 CH_3 groups per 1000 C) affect mostly the rheological properties (viscosity, normal stress) lowering the viscosity as compared to a linear chain having the same DP. Upon decrease of branching, the density rises, sometimes represented by an intermediate group "medium density polyethylene" in the specific mass range of 0.926–0.940. Accordingly the crystallinity as well as the rigidity, tensile strength and melting temperature are all elevated, while the ductility and toughness (impact strength) drop down. These properties are also influenced by the chain length (molecular weight) and distribution. The most prominent properties of polyethylene (representing the whole olefin family that appears as high homologs of various paraffins and waxes) are its chemical inertness towards water, acids and bases. The branched polyethylene LDPE excels also by its low density, high ductility (very high elongation at break) and its high impact strength. As to electrical properties LDPE exhibits a high specific resistivity and a low loss factor. It does not dissolve in organic solvents at temperatures lower than 75°C; at high temperatures the best solvents are the paraffins and olefins. Most mechanical properties drop significantly upon reaching a temperature of 60°C. The elastic modulus and tensile or compressive strength are in the low range, typical for soft polymers. LDPE also suffers from sensitivity to environmental stress cracking (ESC) which appears at loading in contact with polar reagents (mainly detergents). It also shows low resistance to climatic conditions or fire.

It is convenient to augment most properties by an appropriate choice of the polymer grade (high molecular weight and narrow distribution) and proper use of stabilizers and fillers. Regarding ESC a grade of narrow distributed high molecular weight PE should be preferred but alternatives include a blend with synthetic elastomers or a copolymer with butylene. Increasing the crystallinity also elevates tensile strength at the expense of elongation and impact strength. It should be stressed that the producer characterizes the polyethylene grades mainly by the parameters density and melt flow index (MFI).

As MFI appears approximately inverse to molecular weight, a low value is desired for augmenting the mechanical properties. The common range in MFI is 0.1–20. Stabilization against thermal and oxidative decomposition should always be undertaken, while light stabilizers (UVA) and fire retardants are incorporated, when required.

LDPE is easily processed by all common fabrication systems for thermoplastics — injection molding, extrusion, blow molding, rotational molding and thermal forming. Films, pipes, cable and paper coating are accomplished by extrusion. Grades of low MFI are recommended for extrusion and blow molding. Relatively high MFI suits injection and rotational molding (the latter utilizes a powder-like feed). LDPE is easily welded, but difficult to bond. It is also difficult to print (due to lack of polarity) which necessitates an adequate surface treatment (thermal or electrical). A most significant use for LDPE is in films for packaging and agriculture in a thickness range of 0.01–0.25 mm. In food packaging this polymer is advantageous due to its relatively high permeability towards oxygen and carbon dioxide together with its high blocking against water vapor so that it keeps the freshness of the food and enables "breathing." More benefits are its high elasticity, adequate transparency and weldability.

For films in agriculture, the capability of light transmission, the high elasticity and the easy processing enables blowing of thin films at very high width. The minor discrepancy, namely low resistance to solar radiation, can be overcome by means of UV stabilizers. Recently, premium stabilizers for polyolefins (termed HALS) have been offered. When pipes or other accessories are fabricated, they may be excellently protected against weathering by the mere addition of carbon black. LDPE also appears in packaging as bottles and containers (the smaller ones by blow molding while the larger may be rotational molded). Reinforcing fillers may in principle be added to polyethylene, but it is less common when compared to other polymers because of its inertness. Some coupling agents have lately been developed that improve the bonding between polyethylene and the filler. Polyethylene may be expanded (using a chemical or physical foaming process) and cross-linked. The latter is accomplished either by a chemical reaction (mostly peroxides) or by electron radiation.

Cross-linking improves properties at elevated temperatures, rigidity and mechanical strength as well as resistance to ESC. A quite interesting utility lies in the domain of shrinkable films for packaging, while non-cross-linked ones may also be used. There are several modifications. A significant one is based on chlorination so as to gain a structure that is similar to PVC, namely CPE. There are a host of copolymerization processes—with vinyl acetate, EVA; with propylene, EPR (a synthetic rubber); with butene-1 (for enhancing resistance to ESC); with methacrylic acid for obtaining an ionomer (an ionic polymer that will be discussed separately); and many others. In addition LDPE exists in a variety of multiple-layered systems (made by coextrusion), thus reaching the optimum advantage of incorporating material properties. (Much use is made of a 2-layered film of polyethylene and Nylon.) Various blends with EVA and others are also utilized for property enhancement. LDPE has appeared for many years on top of all polymers in production and use, mainly in the fields of packaging, agriculture, piping, electricity and telecommunication, housewares and toys, and adhesives (hot melt). It is also one of the cheapest polymers, showing a distinct correlation between the volume of manufacturing and material cost.

One of the revolutionary novelties in the production of LDPE (apart from the design of gigantic reactors) happened at the end of 1977, when Union Carbide discovered a production method based on low pressure (6–20 atmospheres) and temperature below 100°C. This reactor operates continuously as a fluidized-bed whereas the product is obtained as pellets. The method has a large saving in energy as compared to the old ICI system. Through this method the new type of linear polyethylene with low density LLDPE is produced, as will be explained.

Linear Low Density Polyethylene (LLDPE)   This relatively new polymer in the polyethylene family is essentially a copolymer of ethylene with an olefin group (mostly butene-1, but also hexene or octene), polymerized via coordination. The result is a controlled structure of short branches with a wide range of specific mass 0.910–0.950, the crystallinity and properties varying accordingly. Molecular weights are within $\overline{M}_n = 10^4 - 10^5$. This polymer introduced initially in 1979, now reaches around one third of the total volume consumed of useful polyethylenes. Its production at relatively low pressures, either in the gas phase or in solution, incorporates an energy saving. The new polymer LLDPE exhibits some advantages over its old "brother" LDPE—partly due to its narrower molecular weight distribution $\overline{M}_w/\overline{M}_n < 5$—which produces a distinct elevation in toughness (impact strength) and tear strength. Other advantages include improved thermal properties (melting point). It is much used in extrusion products and shrink-packaging, industrial packaging, agriculture and construction, wherein, one gains properties similar to those of LDPE but at a lower thickness. The degree of crystallinity is higher (65–70%), so that both melting point and stiffness exceed those of LDPE. Due to the narrower MWD, there appears to be less sensitivity to shear conditions while the viscosity is generally above that of LDPE. Both factors somehow hamper the workability. In extrusion film blowing, the stability of the bubble is somewhat diminished and the power requirement increased. Another disadvantage is the appearance of melt fracture. One solution is based on a blend of LDPE with LLDPE, which also compromises the properties of the final product.

Some modifications in the processing machinery are required for LLDPE. Coextrusion of a layer of LLDPE with other polymers, (mainly LDPE, HDPE, PA, PP, EVA), is much performed. The major manufacturers of LLDPE are Union Carbide and Dow Chemical. Recent developments have led to a polyethylene of ultra-low density, in the range of 0.900–0.915, termed VLDPE.

High Density Polyethylene (HDPE)   HDPE has the same chemical structure as LDPE, namely an addition product of ethylene. However, it differs in spatial structure–linear chains with very low branching–leading to high crystallinity (80%–90%), at a specific mass range of 0.940–0.965. It was developed in 1954 as a novel polymerization method, the coordination method based on the use of unique catalysts that determine the structure, all at low pressures. The first inventor of this process (also named stereo–specific) was Dr. Ziegler of Germany. Later, competitive commercial processes were developed by Standard Oil and Phillips in the U.S., the latter comprising at present the major production method for HDPE. The main difference between the three processes lies in the choice of catalyst and its preparation.

The polymerization itself is carried out in a solution in a hydrocarbon solvent in a stirred reactor, at 100°C and pressures up to 30 atmospheres. Contact time is quite long—0.5 to 2 hours. The polymer must be precipitated from the solvent, separated and dried. The linear polyethylene having a high degree of crystallinity differs from the branched polyethylene in the following properties: persistance at higher temperatures (melting at 130°C as compared to 105°C); higher stiffness and tensile strength as well as higher resistance to permeability of gases and to environmental conditions, albeit a higher sensitivity to ESC at similar strains (according to the standard test), and a drop in ductility (lower elongation) and toughness. Light transmission is also reduced with an increase in crystallinity. Accordingly, it complements the use of LDPE in the field of rigid packaging (mainly bottles and containers manufactured via blow molding), boxes, rigid films, rigid pipes, furniture, and electrical insulation. Much use is also made of HDPE in the form of structural foam. Modern uses are as synthetic paper and rigid grocery bags. Similar to LDPE it can be processed by most of the processing methods used for thermoplastics, like extrusion, injection, blow molding, rotational molding and thermal forming.

There exist polyethylenes with an intermediate range of densities, so-called MDPE (specific mass of 0.926–0.940) which enables better performance, like combining resistance to ESC together with rigidity and a barrier to vapors. Currently there are polymerization methods which lead to a broad range of densities at low or high pressure while, at the same time, polyblending of LDPE and HDPE has become popular. One of the interesting novelties is in the use of a linear polyethylene of ultra-high molecular weight (UHM-WPE) wherein molecular weights in the range of 2 to 6 million are obtained. The polymerization process resembles that of conventional ones, but the physical and chemical performances are very distinguished—mainly, tensile strength and rigidity, extremely low friction coefficient, high resistance to abrasion, abatement of noise, and chemical inertness. This type of polyethylene is suitable for very thin packages as well as in the field of engineering for machine parts, medical uses and hot water conduits (after cross-linking). The major difficulty lies in processing because of the high molecular weight which causes extreme high melt viscosity. It is therefore processed by pressing, sintering, extrusion through a plunger, or via pulsation. Manufacturing of sheets and rods is accomplished in a similar method as in metals. Through an appropriate compounding (often by blending with another grade of lower molecular weight PE or a lubricant) and by a special design of processing machines, conventional methods like injection or extrusion can be utilized. Compared to LLDPE, the price of HDPE was higher for many years but there is currently no difference. The use of LDPE is currently more abundant mainly because of fabrication into films for agriculture and packaging. There appears often a competition between the two types of polyethylene in conquering new markets, but meanwhile new polyolefins like LLDPE and PP have been introduced.

Polypropylene (PP) Polypropylene, the third polymer in the polyolefin family, first appeared in 1959, as a result of the development of stereospecific polymerization to obtain ordered polymers of a high degree of crystallinity. Parallel to the invention of Ziegler in Germany, polypropylene was derived by Natta in Italy using similar catalysts. Conditions of polymerization are as follows: pressure of 12 atmospheres, temperature of 30–80°C, mostly as a dispersion. Before that period, polypropylene of the same chemical structure

$$(-CH_2-CH)_n$$
$$\qquad\quad |$$
$$\qquad\quad CH_3$$

was known as an amorphous material lacking any useful mechanical properties. The chemical structure of the monomer indicates a homolog of ethylene, that is, propylene—also a gaseous by-product of oil refineries. In propylene, methyl side groups exist which elevate T_g and give it a higher rigidity and a melting point of 170°C. On the other hand, the appearance of a tertiary carbon in the polymer main chain causes higher sensitivity to oxidation or environmental exposure or at elevated temperatures.

During recent years competitive polymerization methods have appeared, including fluidized bed in the gas phase (similar to LLDPE). In the isotactic structure, polypropylene has 60% to 70% crystallinity, its molecular weight is 25,000 to 500,000 (MFI = 0.5 to 35) and its distribution ratio is 2.5 to 10. Upon increasing molecular weight the yield stress and (mainly) the toughness tend to rise, while ordinarily increasing crystallinity enhances tensile strength and rigidity but diminishes toughness. It is also possible to improve toughness through a special treatment which causes a narrow MWD as well as drop in MW to aid workability. It is lower than polyethylene in density (specific mass of 0.9) which represents a large advantage. Polypropylene has surpassed HDPE in mechanical, chemical and thermal properties including excellent ESC resistance, however, it calls for enhancement of toughness at low temperatures (impact modifiers) and weather resistance. Here it is important to note the development of special stabilizers of the hindered amines (HALS) type. The rigidity and mechanical strength have led to high utilization as special packaging, containers and rigid films, fibers (after stretching for high orientation), rugs, furniture, piping, electronic accessories, transportation (mainly battery cases), machinery parts and sanitary items. The main utility today is as fabric and fibers (mainly for carpeting). Polypropylene is also used in industrial components. Its good chemical resistance at relatively high temperatures introduced polypropylene for tanks in the chemical industry. In general, polypropylene competes mostly with HDPE and PVC (but sometimes with cellophane in oriented packaging OPP). It is already considered a semi-engineering polymer. PP excels in workability mainly by extrusion, injection and blow molding in a temperature range of 230°C to 260°C.

As mentioned before, it is customary to stretch polypropylene uni-axially or bi-axially in order to enhance mechanical properties, including impact strength. It is essential to use antioxidants or UVA for weather resistance. Various fillers and reinforcements are added to PP mostly as fiber glass, talcum or chalk. Copolymerization (mainly as blocks with a minor concentration of ethylene) may be utilized for improvement of toughness. A higher concentration of ethylene in a random copolymer leads to the synthetic elastomer—ethylene-propylene rubber (EPR). Another copolymer, that contains unsaturated dienic groups for cross-linking (EPDM) is also considered as a premium synthetic rubber. It is also possible to improve the toughness of PP at low temperatures by blending with one of the previously mentioned elastomers (while sometimes sacrificing rigidity), by developing processing methods as structural foams, thermal or cold forming. The annual growth rate of the production of PP reached 13% in the 1980s, (already passing that of HDPE) because of its unique properties and its penetration into the market of engineering materials. It also benefited from the petrochemical industry incentive for better exploitation of propylene.

Polybutylene (Polybutene-1)   Polybutylene is the youngest member of the polyolefin family (1965), being linear in structure

$$-(CH_2-CH)_n-$$
$$|$$
$$CH_2$$
$$|$$
$$CH_3$$

The monomer butylene is obtained from the petrochemical industry. Polybutylene is manufactured via a stereospecific mechanism into an isotactic structure, of 50%–55% crystallinity. However, the crystalline morphology shows two modes—a first structure (30%–35% crystallinity) occurs upon cooling from the melt, but quite a different structure is reached through an additional slow crystallization process during about seven days (increasing the crystallinity to 50%–55%). The specific mass of 0.91 is typical of light-weight polymers. MFI appears in the range of 0.4–20. It resembles PE but surpasses the latter in some properties like ESC resistance, relative resistance to creep, and (most of all) it withstands high temperatures (up to 100°C). It is often recommended for manufacturing pipes for the delivery of hot water or for special films. Like LDPE it is ductile and tough with excellent chemical resistance. It may be processed by the same methods as LDPE, including blowing into films. As a new polymer it is more expensive but it is apparently penetrating unique fields where it excels more than polyethylene or polypropylene. Incidentally, a distinction should be made between polybutylene and polyisobutylene—the latter serves as a synthetic rubber, copolymerized with isoprene (butyl rubber).

Polymethylpentene (TPX) This polymer, like polybutylene, is a very young member of the polyolefins (1965). It appears as a polymer based on 4-methylpentene.

$$-(CH_2-CH)_n$$
$$| $$
$$CH_2$$
$$| $$
$$CH$$
$$/ \ $$
$$H_3C \quad CH_3$$

It was developed by ICI using a stereospecific method (Ziegler). It has the lowest specific mass (0.83), high transparency and thermal stability. T_m is 240°C and heat distortion temperature 175°C. Its chemical resistance is similar to that of PE but it may be used at high temperatures. It also needs weather stabilizers. It may be processed by injection or blow molding as well as by extrusion. Due to its low viscosity there appears a strong tendency for orientation during injection—leading to inhomogeneous properties.

It is used for specialties, wherein its transparency and chemical inertness at high temperatures are advantageous—packages that may be boiled, medical facilities, laboratory dishes (replacing glass), lighting and electronics. The production capacity increased mainly since 1973, when royalties were transferred from ICI to Mitsui (Japan), but the price is still high ($2 to $4 per lb).

Polyvinylchloride (PVC) Currently representing one of the four leading polymers (developed in 1936), PVC appears when one hydrogen atom in ethylene is substituted by the element chlorine. The specific mass is 1.4 and it has the chemical structure:

$$(CH_2{-}CH)_n$$
$$\mid$$
$$Cl$$

The monomer VCM is built as $CH_2{=}CHCl$. A typical petrochemical product, PVC is obtained by reacting ethylene (or acetylene) with chlorine or hydrogen chloride, namely chlorination. The presence of chlorine modifies inherently the performance of the polymer (as compared to PE). T_g rises to $+85°C$ (compared to $-120°C$ for LDPE), meaning that PVC is rigid at room temperature, with low elongation and some degree of brittleness—typical for glassy polymers. This is affected by the polarity (dipole moment) which leads to strong secondary bonds. Polymerization is usually carried out via free radical initiation—reaching essentially an atactic configuration with very low crystallinity (about 5%), so it is basically amorphous. The polymer is rigid but easily amenable to becoming plasticized either by external plasticizers or via copolymerization. Therefore, PVC is distinguished by its variety of uses both as a rigid or soft material, the latter state resembling that of elastomers. The industrial polymerization methods mostly in use are suspension, emulsion or bulk. In the aqueous suspension method, there is a good control of the temperature while the polymerization is carried out in a batch reactor (residence time of about two hours) at the temperature of $20°{-}50°C$ and partial vacuum. After separation and drying, a powder is obtained with particle size of $100{-}150\mu$ (frequently in the shape of pearls). By emulsion, minute and very homogeneous particles (in the order of 0.2–2.0 microns) are reached, very suitable for plasticizing and generation of plastisols. Longer chains are also obtained by this method. However, the retention of soap residues hampers the electrical and optical properties. In bulk polymerization, the process is carried out in two stages leading to a polymer with premium physical properties, as a result of the absence of additives. Particle size is in a range similar to that of the suspension method. Following polymerization, the polymer must be separated from the remaining monomer. Since it has been discovered that the monomer VCM may be carcinogenic, the process of separation must be carried out with extreme care, eliminating direct exposure of the workers. A drastic reduction is demanded of monomer content remaining in the polymer (less than 1 ppm), while special restrictions exist for the packaging of liquid food in contact with rigid PVC (to the low limit of 10 ppb monomer in the food).

The polymer characterization deals with the molecular weight and size of the particles. As regards to molecular weight it is customary to use a parameter obtained from relative viscosity measurements of dilute solutions, termed the K-value. The common range is 40 to 70, corresponding to $\overline{M}_n = 20{,}000 - 50{,}000$ and $\overline{M}_w = 50{,}000 - 260{,}000$. The molecular weight distribution is apparently rather narrow. The size and distribution of the particles as well as their structure (mostly shape and porosity) are essential mainly for plastification. All this can be controlled by the appropriate choice of stabilizing agents for the suspension and the mixing velocity. The existence of chlorine and the resultant polarity affect apparent properties such as high chemical resistance, compatibility with many plasticizers and pigments, high dielectric loss factor, good printability and fire retardation due to the elimination of HCl. On the other hand, PVC is supersensitive to heat and care must be taken during processing to use relatively low temperatures to eliminate overheating. Therefore, machinery for the processing of rigid PVC is specially designed, displaying deep channels and lack of metering zones in extrusion, generous design of the dies (elimination of sharp corners), anticorrosion protection of the metal towards acidic HCl, and a shortened thermal history in injection molding. An extended period above a temperature of 200°C must be avoided, albeit the processing of rigid PVC is carried out at this range. Special thermal stabilizers are applied in addition to light stabilizers and, incidentally, the use of twin extruders has been developed during the processing of rigid PVC. Stabilized PVC bears excellent light transmittance, whereas the maximum temperature in use is 60°C. PVC can also be protected by appropriate pigmentation. It dissolves in few organic solvents, but plasticizes (swells) in contact with many stabilizers. The addition of plasticizers (up to 50% in weight) largely improves the workability, but the main goal is to achieve a soft and ductile material (elastomeric) while the molecules of the plasticizer align themselves between the chains (a solid solution) and thus lower their secondary bonds. The hard and rigid polymer turns flexible and soft, with high elongation and ductility. With the addition of plasticizer, the toughness at low temperature is enhanced, as a result of lowering the T_g. Special impact modifiers are added to rigid PVC, mostly of the type EVA, ABS, acrylates or chlorinated polyethylene.

In most cases, the first stage in compounding is comprised of dry blending or melting of the resin powder with all the additives, followed by granulation in an extruder in order to obtain even pellets. Rigid PVC (minimal plastification below 10%) may be processed at temperatures of 150°–185°C, with great care taken as to processing conditions and appropriate machine design, via callendering, extrusion (pipes, profiles and films), injection or blow molding, compression, thermal (vacuum) forming and rotational molding. Plasticized PVC is mainly processed via callendering, extrusion (including insulation of electrical wires), injection and rotational molding (the latter mostly as plastisols). Plastisols may also be formed by casting, dipping or coating. Both rigid and soft PVC may be foamed in various modes. The use of these foams is becoming significant—structural foam of rigid PVC for furniture, or soft foam for upholstery including "sky" from plastisols that resembles leather. PVC also appears as latex (in emulsion), which may be fabricated like rubber lattices. The major breakthrough is in the field of blow molding of bottles from rigid PVC, and the adaptivity of the twin extruder or modern injection machines for efficient and precise forming of rigid PVC.

Items made of PVC may be bonded through welding (hot air) or by glues and cements based on the same materials (like in bonding of pipe systems). In case of thin films it is convenient to use dielectric heating at high frequency which is mainly suitable for PVC due to its high electrical loss factor. As mentioned before, PVC forms an extensive and versatile compound, including stabilizers, lubricants, plasticizers, fillers, pigments and other additives. Accordingly, a broad range of physical properties is achieved so that it is difficult to tabulate the ultimate properties of plasticized PVC. As a result, rigid PVC appears in versatile uses as for stiff pipelines (for water supply, gas, sewage and electrical conduits), sanitary accessories, sheets for partitions and roofs, shutters, windows and doors, containers and structures in the chemical industry, packaging and bottles. There is increased use in construction, including prefabricated buildings, but the danger of creep-under-load must be taken into consideration. It can be supplied as transparent, translucent or opaque PVC in a variety of colors. Plasticized PVC serves in the insulation of cables, irrigation pipeline, flooring, conveying belts, food packaging, records, upholstery, shoes, wall and container coating, paints, cloths, toys and sports equipment. The variety of uses has made it almost indispensable. Currently, polypropylene competes with PVC due to similarity in properties and markets (including costs). In various applications (as in agriculture, packaging or insulation of electrical wires), plasticized PVC competes with polyethylene, in spite of the higher cost of the former. PVC shows several advantages over PE. It is more stable towards the environment (but needs stabilization), is less permeable to vapors and gases in films, it does not suffer from ESC and in general it is easy to compound with many additives. On the other hand, there appears a danger of migration of the plasticizer when exposed to the sun, thus turning the soft and ductile material into a rigid and brittle one. PVC plastisol solidifies after heating, and is easily processed by simple methods (also by foaming) obtaining an underlayer for rugs, textiles, toys and life belts. One interesting combination consists of a polyblend of PVC with ABS.

Another modification is reached by adding chlorine—chlorinated PVC or CPVC—which serves as pipeline for hot water supply, but is also very sensitive to thermal degradation. Various copolymers are based on vinyl chloride with a second component, mostly, being vinylidene chloride $CH_2=C\,Cl_2$ (called also Saran) or vinyl acetate:

$$H_3C-\overset{\overset{\displaystyle O}{\|}}{C}-O-\overset{\overset{\displaystyle H}{|}}{C}=CH_2$$

In both cases, an internal plastification is achieved with no danger of migration of the plasticizer. Saran is much used in clothing and textiles, food packaging and paints as well as in rigid structures. The copolymer with vinyl acetate is mostly used for records.

Polyvinylidene chloride appears as a crystalline polymer excelling in chemical resistance and low permeability, but it is difficult to process and thermally unstable. Both homopolymers vinyl acetate and vinyl alcohol appear as useful polymers, the former in emulsion (coatings and glues) and the latter in various coatings (colloids and additives in textiles) and in food packaging. Other vinyl polymers are polyvinyl acetal and butyral. In conclusion, polyvinylchloride is a well established polymer with a broad range of properties and uses. The major drawbacks of the rigid type are thermal instability and sensitivity to processing conditions. Currently there is some hesitancy in using rigid PVC (mainly in contact with food and water) due to the danger of possibly eliminating the monomer (though this issue can be overcome at a price). In any case, the future of rigid PVC is not clear, taking into account difficulties of direct recycling or reuse as a source of energy.

Chlorinated Polyether (Penton)

$$-(H_2C-\underset{\underset{CH_2Cl}{|}}{\overset{\overset{CH_2Cl}{|}}{C}}-CH_2-O)_n-$$

This is a special crystalline polymer (melting point 181°C with specific mass equal to 1.4) which has been manufactured since 1959. Its advantages appear in thermal and chemical properties surpassing those of rigid PVC. This performance places it between PVC and fluorocarbon polymers. Due to the location of the chlorine atom, it does not decompose like PVC, but is self extinguishing. The polymer excels in rigidity, dimensional stability and endurance under corrosive conditions. Consequently it is categorized as an engineering polymer, being used for pipelines, pump accessories and valves that withstand chemicals at temperatures up to 125°C. It may be processed by injection molding and extrusion. It is stable at processing temperatures of 180°C–240°C and the heat distortion temperature of 140°C.

Fluorocarbon Polymers (Fluoroplastics)   These represent a broad family of polymers based on hydrocarbons containing fluorine, and sometimes also chlorine. The most distinguished polymer in this group is polytetrafluoroethylene (PTFE), known more by its commercial name of Teflon™ (1943). In this case, all four hydrogen atoms in ethylene are replaced by fluorine in a linear chain. The specific mass is 2.15.

$$(-\underset{\underset{F}{|}}{\overset{\overset{F}{|}}{C}}-\underset{\underset{F}{|}}{\overset{\overset{F}{|}}{C}}-)_n$$

The presence of fluorine contributes extraordinary properties such as superb thermal and chemical stability and low friction coefficient. Replacement of fewer hydrogens or a combination of chlorine and fluorine leads to polymers that are somewhat inferior to Teflon, but still much better than most thermo-plastics.

$$\begin{array}{ccc}
\text{H}\;\;\text{H} & \text{H}\;\;\text{F} & \text{F}\;\;\text{F}\\
\mid\;\;\;\mid & \mid\;\;\;\mid & \mid\;\;\;\mid\\
(-\text{C}-\text{C}-)\text{n} & (-\text{C}-\text{C}-)\text{n} & (-\text{C}-\text{C}-)\text{n}\\
\mid\;\;\;\mid & \mid\;\;\;\mid & \mid\;\;\;\mid\\
\text{H}\;\;\text{F} & \text{H}\;\;\text{F} & \text{Cl}\;\;\text{F}
\end{array}$$

PVF	PVDF	PCTFE
(polyvinyl-fluoride)	(polyvinylidene-fluoride)	(polychloro-trifluoro-ethylene)

The monomer tetra-fluoroethylene is obtained by reacting HF with chloro-form through cracking, whereas the source for HF is fluorspar CaF_2, treated with sulfuric acid.

All other polymers in this family are also obtained by addition polymeri-zation of the appropriate monomers (or pairs in copolymers), but they are more expensive than conventional polymers. The ordered chemical structure of PTFE (spiral-like) leads to high crystallinity, and due to the high bond strength of C–F (120 kcal/mole) the resulting melting point is very elevated, 327 °C, while most mechanical properties are retained close to this tempera-ture. Secondary bonds are not high (lack of polarity) so that this polymer is not rigid, its strength is moderate and it suffers from a significant creep. In practice, it is very difficult to melt Teflon^{TM}, and there appear to be no good solvents except for specialties at the melting point.

Teflon^{TM} appears with ultra high molecular weights $\overline{M}_n = 4 \times 10^5 - 9 \times 10^6$, whereas crystallinity rises (up to 94%) upon decreasing MW. In disper-sion polymerization, lower molecular weights are obtained with smaller par-ticles—resulting in improved mechanical properties. Teflon^{TM} is a ductile and tough polymer, excelling in endurance at high temperatures, and showing extreme inertness. It is not attacked by most chemicals, insensitive to water, inert to the atmosphere, non-burning, does not stick and displays an ex-tremely low friction coefficient. It shows an amazingly wide temperature range, between −269 °C and +260 °C throughout which it preserves its flexi-bility. In addition it excels in electrical properties, including dielectric strength. In order to exploit these special properties there is a significant price—both in cost of the resin, which happens to be one of the highest, and in the requirement of specific processing methods. Because it does not melt, it may be formed solely by cold pressing of the powder or by sintering similar to ceramics and metals. Under special conditions it is possible to extrude by means of a plunger into rods and tubes from which the final shape can be formed by milling. Quite another type of extrusion is performed as a paste upon the addition of volatile plasticizers. Compression molding is also used, while dispersions are sprayed for coatings. (Other fluorocarbons are easier to process, but fail in properties.) It is usually not amenable to welding, but bonding may be achieved with special glues. The problem of creep may be partly overcome by the aid of special fillers. The unique uses of Teflon are seals, rings and bearings, and coatings for accessories that withstand high temperatures and corrosion. It is much used for sealing-bands, electrical insulation in the region of high frequency and temperature, and antistick coatings for cooking vessels or shaving razors. The nearest fluoro-polymer in properties to those of Teflon, is the copolymer fluoro-ethylene-propylene (FEP).

$$CF_2-CF_2: CF_2-CF$$
$$|$$
$$CF_3$$

It can advantageously be injected, extruded, compressed or thermally formed. It also surpasses Teflon in rigidity and resistance to creep, but relatively fails in thermal and chemical endurance, being useful up to 200 °C.

PCTFE also is distinguished by its easy workability in regular machinery. It is used as thin transparent film in sterile medical packaging (via radiation), due to its excellent resistance to outdoor radiation and water permeability. It is also found in various engineering and electronic applications. PVDF (Kynar) can also be processed into rigid tubes for the chemical industry enduring a wide range of temperatures, -80 °C to $+300$ °C. However its chemical resistance is inferior to that of Teflon^{TM}. It is also much used as a coating. PVF (Tedlar) appears mainly as flexible film, excelling in weather and chemical resistance, but it is relatively unstable during processing (like PVC). Its useful temperature range is -100 °C to $+150$ °C. It may be used for the coating of wood, metal and composites in the aviation industry.

There are more polymers in this family, including copolymers (with ethylene in alternation). Upon replacing atoms of fluorine with chlorine or hydrogen, the chemical inertness and the range of useful temperatures diminish, while workability, strength and rigidity all improve. The price of these polymers is quite high, so they are used only when their unique performance is required.

Polystyrene (PS)   PS is an old (1938) and very useful polymer, but its production capacity has been diminished compared to other commodity thermoplastics. It is a linear polymer with specific mass = 1.04.

$$-(CH_2-CH)-$$
$$|$$
$$\bigcirc$$

It represents a vinyl in which one atom of hydrogen is substituted by a benzene ring in an atactic structure (no steric order). Therefore it is completely amorphous, glassy and brittle, because the bulky ring hinders compact ordering. T_g is also elevated to 100 °C. It acquires high transparency, rigidity and easy workability. It appears in several grades—general purpose grade (GP) without any modification, high-impact polystyrene (HIPS) built as a copolymer with butadiene, expandable polystyrene (EPS), or modified polystyrene for improved flow (plasticizers). The monomer styrene is one of the cheapest, and also appears in other polymers and copolymers in the field of elastomers and unsaturated polyester. It is obtained as a by-product of crude oil refining, through activation of ethylene on benzene (ethyl-benzene appears as intermediate). It is a liquid boiling at $+145°C$, sensitive to heat and light as it easily tends to polymerize, so that an inhibitor should be added. Styrene polymerizes in an addition reaction (usually radical) via any of the following methods—bulk, solution, suspension and emulsion—either batchwise or continuous. In the bulk process (80°C to 180°C) a pure and transparent polymer is achieved with excellent electrical properties. However, in order to achieve a better temperature control and elimination of a wide distribution, the methods of suspension and emulsion are preferred, which eventually require separation and drying. The mean molecular weight is 200,000. General purpose polystyrene (GP) is rigid and brittle, with distinguished optical and electrical performance, excellent workability, however it is sensitive to the environment (due to the release of monomer residues that cause yellowing). It withstands many chemicals and water, but is sensitive to some oils and organic solvents (including crude oil) and tends to crack under stress. It is limited to working temperatures of 60°C to 80°C. It may be processed by injection to high accuracy, by thermal forming, extrusion and blow molding.

Polystyrene also appears as structural foam. In order to overcome the major discrepancy, its brittleness, the HIPS grade has become more dominant. In the past, a blend with a synthetic rubber (at a concentration of a few percent) was used, wherein the soft elastomeric phase stops crack propagation. Currently, this is achieved by copolymerizing with polybutadiene (about 4% to 8%) through a complex process ending with suspension polymerization. The process is based on grafting of styrene to a polybutadiene chain and then cross-linking of the double bond in butadiene. The concentration and size of the rubber sphere particles (in the range of 0.5 to 5 microns) distributed as a separate phase in the free polystyrene, determine the impact strength. Hence there are several grades of HIPS—the super one enhancing impact by up to eight-fold—but upon increasing toughness, the transparency, rigidity and tensile strength tend to drop. Other modes for enhancement of toughness are based on biaxial orientation (also improves weatherability), as well as by foaming, where a volatile liquid (pentane) is introduced to the polymeric system through thermal processing. The expandable compound is supplied by the producer of raw material as granules.

The scope of utility of PS is very broad, and HIPS serves as machine bodies, radio and television cabinets, food packaging and dishes or refrigerator doors. The structural foam is found in furniture, containers and engineering structures. Foamed PS appears as a rigid and tough material in containers and packaging, water vessels and plant pots as well as insulation boards in construction. GP grade is still used for toys, household dishes and cheap packaging, but has been mostly replaced by HIPS. PS requires fire protection via special additives. Major efforts are devoted to the development of PS grades that combine toughness with strength and transparency, and to the search for other types of elastomers that will improve weatherability. Fire retardancy too should be improved by use of active elements. It is possible to obtain PS with a very narrow MWD via an anionic polymerization (living polymers), but its use is trivial and mainly for research or calibration. A method for manufacturing crystalline isotactic PS has also been developed, but the polymer suffers from brittleness and poor workability. Many copolymers based on styrene are known, the most prominent one being the synthetic rubber polybutadiene–styrene – about 25% styrene – that also serves in coatings. Among the plastics, one should mention the copolymer styrene–acrylonitrile (SAN) that contains about 25% acrylonitrile.

$$-(CH_2-CH)-$$
$$\underset{\displaystyle CN}{\vert}$$

SAN gives improved toughness, resistance to solvents and crazing, good thermal properties, while retaining the transparency of the polymer. Upon increasing the concentration of acrylonitrile in styrene copolymer some properties are enhanced except for workability. SAN is more improved in properties than PS and hence competes (in price or performance) with acrylics and cellulose-acetate (except for weatherability).

Another improvement in the mechanical properties (toughness, rigidity and strength) is reached by synthesizing the terpolymer ABS, wherein a third elastomeric component (butadiene) is added to acrylonitrile–styrene. This terpolymer has been acknowledged and esteemed as a semi-engineering material, that deserves its own description.

Acrylonitrile–Butadiene–Styrene (ABS)   There are about 15 different grades of this terpolymer varying in composition and mode of manufacture. This unique terpolymer, developed in 1946, is composed of the monomers acrylonitrile, butadiene and styrene in various concentrations. The most common grade is manufactured either by grafting the copolymer styrene–acrylonitrile (SAN) on a backbone of butadiene rubber, or by blending SAN with nitrile-rubber (copolymer of butadiene and acrylonitrile). As mentioned before, various combinations may appear that determine the ultimate performance, mainly the balance between three factors – rigidity, strength and toughness. The advantage of ABS over HIPS lies mainly in an increase of impact strength and better thermal or chemical properties without reducing rigidity or tensile strength. ABS is easily processed via injection, extrusion, blow molding, thermal forming, as well as cold pressing, metallic coating and structural foam. It is also easily welded by ultrasonic methods or glues. ABS can be manufactured in bulk, suspension or emulsion, frequently with the in situ addition of special additives for fire retardancy, foaming, etc.

In contrast to SAN, it is not transparent and needs stabilization for external exposure. The price of ABS is about 1.5 times that of PS, and it suffers strong competition from PP, which is also much cheaper. However, ABS is considered more engineering-like. Its range of utility is wide – pipelines for gas and chemicals, cabinets for radio or TV, telephone sets, batteries, machine bodies, luggage, helmets, sport accessories, boats, car bodies and machinery. In novel developments, it appears with α-methyl styrene together with styrene (as a fourth component), in polyblend with PVC or polycarbonate, and with the incorporation of reactive fire extinguishing elements. Another terpolymer MBS (methacrylate–butadiene–styrene) has been developed which is transparent and has good mechanical properties.

Acrylics   These are a versatile family of polymers, most commonly represented by polymethyl methacrylate (PMMA), or commercial names, Perspex or Flexiglass. It was developed in 1931. Its specific mass $= 1.18$. The chemical structure is as follows:

$$-(H_2C-\underset{\underset{COO\,CH_3}{|}}{\overset{\overset{CH_3}{|}}{C}})_n-$$

With time, more acrylic polymers and copolymers have been developed, mainly acrylates, methacrylates and acrylonitrile. PMMA is distinguished as a glassy polymer, $(T_g = 120\,°C)$, rigid and hard, excelling in transparency and high weatherability, therefore it appears as a distinct replacement to glass. The monomer is obtained through a reaction including acetone, hydrogen cyanide, sulfuric acid and methanol. It is more expensive than the monomers described so far. The common polymerization mechanism is radical in bulk (mostly by casting into molds) or in suspension. Like styrene it polymerizes easily, so that the monomer must be inhibited during storage. Bulk polymerization began in the 1930s and still is used abundantly for sheets, rods and tubes. The liquid monomer is directly cast into a mold, with catalysts and a small amount of polymer in solution in order to obtain a viscous syrup with low shrinkage. The mold itself consists of two smooth glass walls, held at a temperature of 40°C to 100°C. The batchwise process takes about 15 hours, until very high molecular weights (around 10^6) are reached. There are also continuous processes, and sheets can be obtained by extrusion but their optical properties are relatively inferior. Suspension polymerization was developed later (residence time around one hour). Through this process lower molecular weights are reached (about 60,000) which are advantageous for injection molding and extrusion. Like PVC, "pearls" are attained in controlled sizes, but they can also be pelletized. The chemical structure, in which a polar ester group and a methyl branch exist in a random or atactic configuration, leads to a glassy amorphous state with high rigidity. It is sensitive to organic solvents, mostly aromatic and polar (like acetone). The sheets exhibit high light transparency and weather resistance, but residual stresses should be annealed while biaxial stretching and partial cross-linking are desirable for overcoming the tendency to craze. The working temperature is up to 95°C.

The sheets can be thermally formed under pressure or vacuum, and in addition can be bonded with an appropriate solvent or cement. Granulates, pigmented or unpigmented (translucent to opaque), are easily shaped by injection molding or extrusion, albeit their higher melt viscosity (compared to PS) calling for higher pressures or temperatures. Machine milling is also popular, starting with sheets, rods and tubes.

As for impact strength PMMA surpasses polystyrene or glass, but is susceptible to scratching. Compared to glass it is safer upon breaking, so that it serves as a substitute for glass wherever the high price is not prohibitive — windows, partitions, ceilings and domes in constructions or aviation, lighting, lenses and glasses, and greenhouses. Tinted PMMA (extraordinary high gloss) is very popular in sanitary units like baths and sinks, as well as a replacement for tinted glass in art works. Other uses involve dentistry, artificial limbs, and jewelry. Various copolymers of acrylates are based on methyl-methacrylate with methyl-acrylate, ethyl-acrylate, acrylonitrile or styrene. Those copolymers are useful in paints and varnishes or for optical equipment (including contact lenses). Methyl-methacrylate also replaces styrene as a cross-linking monomer with unsaturated polyester for improved weatherability. Acrylic films may be utilized as the upper layer in a laminate with other materials for enhancement of weather resistance. A rubber modified grade elevates the toughness, thus approaching the mechanical properties of polycarbonate (which competes successfully as a transparent engineering polymer with PMMA, in spite of its higher cost). Another modification is based on blending with PVC for enhancement of fire retardancy. Acrylonitrile appears in many copolymers. In Barex (1968) it was used in bottles that show low permeability to gases, thus being able to replace glass for gaseous beverages; however, this use has been banned due to its toxicity (suspicion of carcinogenicity). A well established use of acrylonitrile is in the synthetic rubber NBR (with butadiene) that exhibits excellent oil resistance, as well as in acrylic textiles, like Orlon. Its appearance in ABS (together with butadiene and styrene) has already been described.

Acetal, (Polyacetal) Poly-oxymethylene (POM) Acetal is a polymer obtained through an addition reaction of formaldehyde $-(CH_2-O)_n$. It excels in mechanical performance and is regarded as a prominent engineering polymer. It appeared in 1959 with the commercial name Delrin™. A short time later a useful copolymer was also developed with a cyclic ether like ethylene oxide. The monomer formaldehyde is a gas produced mostly by oxidizing methanol, and it is very useful in thermoset polymers like phenol, urea and melamine–formaldehydes. For high purity it is initially converted to trioxane or para-formaldehyde. The polymerization is carried out by ionic mechanism, wherein the monomer is dispersed in an inert liquid (heptane). The molecular weights reach 20,000 to 110,000.

If the structure of acetal is compared to that of polyethylene, it is apparent that by replacing the methylene group (CH_2) with an oxygen atom, a compact structure is reached leading to high crystallinity (about 80%) and melting

point around 160°C. It is also relatively flexible due to the flexible bond (CO),
so that its T_g is rather low at (-78°C). The resultant combination is unique –
a polymer with high mechanical strength, rigid but also tough, with unusual
springlike properties, dimensional stability, low creep and friction coefficient, high resistance to abrasion and fatigue, stability towards solvents and
chemicals, and a broad working temperature range -50°C to $+160$°C. However, the polymer should be protected (when exposed to the atmosphere)
against UV radiation. It is no surprise that acetal appears as the foremost
engineering polymer replacing metals in typical construction uses like bearings, gears, springs, handles, transportation and sanitary accessories, bodies
of pumps, electronics and agriculture.

Both homopolymer and copolymer may be processed by injection molding
(the main method), extrusion and blow molding. In the latter case, the copolymer has the advantage of higher melt strength. The copolymer is less susceptible to decomposition to formaldehyde, and more resistant to hydrolysis
when exposed to the environment, but the homopolymer has better mechanical properties. Acetal can easily be welded but is less easily glued. In spite of
its relatively high cost, acetal appears as a prominent engineering polymer
and is especially used with fiber reinforcement.

Polyamides (PA), Nylon   Polyamides represent a family of polymers based
on the amide group

$$-\overset{\displaystyle H}{\underset{\displaystyle H}{N}}-\overset{\displaystyle O}{\overset{\|}{C}}-$$

The name Nylon was given by the DuPont Company which produced the
first polyamide Nylon 6-6 in 1939. It caused a real revolution in the world of
textiles, when the first premium man-made fiber was introduced. It is not
surprising that the nickname Nylon has been used for any polymer by many
people (due to lack of sufficient knowledge). In most cases these polymers are
manufactured by a condensation reaction (stepwise) between a diamine and
di-acid, wherein the numbering indicates the number of carbons in the amine
and acid, respectively. So, the original Nylon 6-6 developed by Carothers, is a
product of the reaction between hexa-methylene diamine and adipic acid.
Nylon 6 was developed in Germany (1943), via a process of ring opening of
caprolactam. Since then many other types of Nylon have been developed, the
most important ones being 6-6, 6, 6-10, 6-12, 11, 12. See Table 6-1.

The typical reaction for the production of Nylon 6-6 is as follows:

$$\text{n HOOC(CH}_2)_4\text{COOH} + \text{n H}_2\text{N(CH}_2)_6\,\text{NH}_2 \rightarrow$$
$$\text{H}-[\text{HN (CH}_2)_6\text{NHCO(CH}_2)_4\text{CO}]_n-\text{OH} + (2\text{n}-1)\,\text{H}_2\text{O} \qquad (6\text{-}1)$$

TABLE 6-1
Structure of Commercial Nylons

Generic Name	Specific Mass	Di-amine	Di-acid	Single Monomer
Nylon 6	1.14	—	—	$H_2N-(CH_2)_5-\overset{\overset{\displaystyle O}{\|}}{C}-OH$ ω-amino-caproic acid $\underset{HN-CO}{(CH_2)_5}$ caprolactam
Nylon 6-6	1.14	$H_2N(CH_2)_6NH_2$ hexamethylene diamine	$HOOC(CH_2)_4COOH$ adipic acid	—
Nylon 6-10	1.08	$H_2N(CH_2)_6NH_2$ hexa-methylene diamine	$HOOC(CH_2)_8COOH$ sebasic acid	—
Nylon 11	1.05	—	—	$H_2N(CH_2)_{10}COOH$ ω-amino-undecanoic acid
Nylon 12	1.02	—	—	$\underset{NH}{(CH_2)_{11}-CO}$ dodecyl(lauryl) lactam

In this condensation reaction, water is eliminated. The structure of Nylon 6 is

$$H-[HN (CH_2)_5-CO]_n-OH$$

The monomer for Nylon 6-6, adipic acid, is derived from benzene or phenol (aromatic petrochemicals) after a series of hydrogenation and oxidation reactions. The other monomer, hexa-methylene diamine is derived from several sources, but may also be obtained from adipic acid itself. It is interesting to note that when Nylon 6-6 was initially produced in 1935, this monomer was not yet commercially used.

The main issue in condensation polymerization is the retention of equivalent concentrations of both contra-functional groups (COOH and NH_2), because an excess of either one eventually decreases the chain length significantly. Hence it is customary to react both monomers in a first stage forming a Nylon salt (hexa-methylene diammonium adipate) as an aqueous solution which is conveniently stored or transported. Polymerization is carried out in a stirred tank (autoclave), either batchwise or continuously, at pressures up to 20 atmospheres and temperature of 200°C to 275°C.

In the final stages, partial vacuum is applied in order to eliminate the condensate water. A batch process takes about 5 to 6 hours. Molecular weights of the order 10,000 to 25,000 are obtained, where a regulator is used to control the chain dimensions (a monofunctional group that attacks the reactive end). It is amazing now to remember that when Carothers reached molecular weight of 10,000 in his work, he referred to it as a superpolymer. Today the range of superpolymers lies in the millions, but with condensation polymerization it has already been explained that one cannot reach ultra high molecular weights which are not favored for the fiber industry.

The molten polymer exudes through an extruder to produce pellets or thin fibers, followed by stretching, in the spinning process. The unique properties of Nylon are attributed to the existence of the amide group (−CO−NH−) at a constant frequency, leading to polarity and strong hydrogen bonds. As a result, a high crystalline compact structure is obtained, with high melting points, high mechanical strength and rigidity, as well as toughness related to the rather long methylene groups. The longer the distance between the amide groups (like the transition from Nylon 6-6 to Nylon 11 or 12) a more polyethylene-like structure is obtained. The advantage of types 11 and 12 are that in spite of a drop in melting point and strength, ductility and resistance to water absorption are improved. Due to the low molecular weights, the melt viscosity is also low. The high crystallinity renders Nylon poorly soluble in organic solvents.

Nylon, beside its important utility as premium fiber in the textile industry, also serves as an engineering polymer due to its unique properties of rigidity and toughness, low friction coefficient (including self lubrication), high resistance to abrasion and fatigue, supreme chemical resistance (including fire retardancy), as well as excellent thermal and electrical performance.

The high melting point of Nylon 6-6 (260°C) and $T_g = 45$°C, attribute superb thermal properties and a working temperature of 150°C. On the other hand, if no special protection is applied, Nylon because of its structure suffers from sensitivity to humidity (about 8% water absorption for Nylon 6-6), and to UV radiation and oxidation by the environment.

When Nylon is processed via extrusion, injection or blow molding, the screw should be designed with a sudden transfer to compression because of low viscosity and short melting range. Feed must be dried to 0.3% humidity and stored under dry conditions. The pressure in injection is also relatively low. Other modes of processing are by sintering, powder coating, or rotational molding of Nylon 6 from the monomer during polymerization.

For engineering purposes the higher range of molecular weight is preferred as compared to fibers. During cooling of the molds, care should be taken to avoid loss of crystallinity (50% to 60% crystallinity for moderate cooling as compared to 10% for quenching). Although Nylon 6 competes with Nylon 6-6 mainly in Europe (due to lower price and improved workability albeit inferior chemical and thermal performance), both types are utilized in the engineering market for machine accessories (mostly with fiber reinforcement), gears and bearings, pump bodies, special containers, medicine and transportation, handles, brushes, rugs, electrical or mechanical tools, tubing and special valves. All this is in addition to the fiber industry (textiles, ropes, fishing rods, tennis rackets) or for packaging food as barrier films. The resistance to various chemicals, including oils and solvents, is superb, with the exception of concentrated acids that cause hydrolysis. However, this deficiency may be exploited for recovery of raw material from the waste.

Humidity affects the strength and rigidity while also reducing electrical properties, but it serves as a plasticizer and enhances toughness. The novel types like 6–10, 11 and 12 are definitely more expensive, but they are unique due to lower sensitivity to water as well as improved chemical resistance. The water repellency increases with the length of the methylene groups. There are also several copolymers, some within the family, like 6–6/6 or 6–6/6–10. Other Nylons™ of limited utility are termed 4, 7, 8 or 9. Novel developments are mainly in the aromatic polyamides like Kevlar

$$- \bigcirc -CO-NH- \bigcirc -NH-CO-$$

which excels at high temperatures and serves as a superb fiber for tires or general reinforcement.

Another aromatic polyamide, Nomex, is suitable for nonburning textiles.

$$- \bigcirc - \overset{\overset{\displaystyle H}{|}}{N} - \overset{\overset{\displaystyle O}{\|}}{C} - \bigcirc -$$

The main polyamides (Nylon 6 and 6-6) are the oldest and most prominent representatives of engineering polymers, mostly with glass fiber or other fiber reinforcement. Improved mechanical performance and a relatively high temperature range of use are achieved because of the crystalline structure.

Saturated Polyesters (PET, PBT) The most prominent polymer in this family is polyethylene-terephthalate (PET), developed in 1945 as a condensation product between terephthalic acid and glycol. This linear polyester is used as a fiber (Terylene™ or Dacron™) or a film (Mylar™), but its main utility lies in the domain of bottles for soft drinks. Polybutylene-terephthalate (PBT) has also been introduced as an engineering material in addition to textiles.

PET is produced from teraphthalic acid and ethylene glycol, while PBT is made with butylene diol. The diacid is obtained by catalytic oxidation of para-xylene (a by-product of the petrochemical industry) while ethylene glycol is obtained from ethylene oxide (an oxidized derivative of ethylene). The terephthallic acid is preferably replaced by the ester dimethyl-terephthalate

$$n\ HOOC- \bigcirc -COOH + nHO- CH_2 -CH_2 -OH \rightarrow$$

$$HO\ (CH_2 -CH_2 -OOC - \bigcirc -COO)n\ H + (2n-1)\ H_2 O \qquad (6\text{-}2)$$

Other polyesters are based on a reaction between glycols with various paraffinic acids (adipic or sebasic) or the polymerization of a hydroxy-acid or lactone (a single monomer) resembling hydroxy amine or lactam, in the synthesis of polyamides. The processes are similar to the polymerization of polyamides. Like the Nylons, the saturated polyesters are used in fibers or engineering plastics (magnetic tapes). In the latter case higher molecular weight grades are preferred. Polyester is relatively new in the field of engineering polymers, PET since 1966 and PBT since 1974. In spite of its higher cost, PBT has several advantages such as rigidity, dimensional stability at elevated temperatures, low water absorbency, low creep, improved chemical resistance, as well as better processability (compared to PET). PET is a crystalline polymer (over 40%), with some qualities that exceed those of Nylon, $T_m = 265°C$ and $T_g = 80°C$. It is more rigid and stable than Nylon, but inferior in tensile strength and toughness. The average molecular weight (number average) is around 20,000.

The major processing method is by injection molding or casting. By a novel method developed in the late 1960s—blow molding with in situ biaxial stretching above T_g—PET gains toughness and transparency with barrier properties towards vapors. Consequently, PET gained from this revolutionary process a significant advantage in containers for soft gaseous beverages. The tremendous increase in use of PET rendered it a popular commodity polymer. A modern application is in trays heated by microwave.

One deficiency of PET lies in the requirement for a hot mold (140°C) for optimal crystallization during injection, which in spite of higher cost shifted use towards PBT in injected items for engineering. By reinforcing with glass fibers (30%) the heat distortion temperature of PBT rises from 70°C to 210°C, an amazing achievement in addition to rigidity and strength, dimensional stability and superb durability. It is much used in transportation, electrical accessories, household and machinery. Needless to say, as a relatively new engineering polymer, polyester has to compete with older ones, like Nylon and polyacetal. Special polyesters withstand extreme temperatures (HT), like Ekonol—an aromatic polyester, that endures temperatures of 250° to 300°C.

The unsaturated polyesters (mainly used for coatings, including alkyds) are probably better known. However this type of polyester will be discussed later together with other thermosetting polymers.

Polycarbonate (PC)   This polymer, a branch of polyester, was developed in 1958, its main structure being bis-phenol A

$$H \{O-\bigcirc -\overset{\overset{\displaystyle CH_3}{|}}{\underset{\underset{\displaystyle CH_3}{|}}{C}}- \bigcirc -O-\overset{\overset{\displaystyle O}{||}}{C}\}_n-H$$

The molecular weights are in the range $\overline{M}_n$ = 30,000–50,000. It is obtained through a condensation reaction between phosgene $COCl_2$ and di-hydroxy-bisphenol A. The unique bis-phenol A structure that together with the aromatic rings contains the carbonate radical

$$\underset{\underset{\displaystyle O}{||}}{O-C-O}$$

has attributes of high stiffness and strength with a relatively high heat distortion temperature (135°C), which is uncommon with unreinforced thermoplastics (except for Teflon). In addition PC has a high T_g in the proximity of 150°C, and yet this polymer is not brittle like polystyrene. On the contrary, it exhibits an extremely high toughness, probably due to the appearance of an additional transition temperature, $T_\gamma = -100°C$, related to the rotation of the carbonyl radical. While it is usually amorphous, it also excels in high transparency, though it may be crystallized at elevated temperature. Consequently, the rare combination of good optical, thermal, mechanical and electrical properties (it is also self-extinguishing), makes PC a supreme and widely demanded material in the domain of engineering polymers. However it also suffers from some deficiencies, such as sensitivity to humidity (special care must be taken during processing) and to some chemicals, to ultraviolet radiation (requiring stabilization), as well as to crazing under stress. It may be processed by the conventional methods like extrusion, injection molding, blow molding or thermal forming. It is also processed as structural foam or fiber reinforced. It is necessary to dry the resin prior to processing, otherwise it tends to yellow and create bubbles. It is processed at rather high temperatures (230° to 300°C), while the melt viscosity is relatively high and less sensitive to shearing (almost Newtonian). Annealing is recommended at temperatures above 135°C. It is useful for such things as lighting fixtures, electronics, photographic films, unbreakable dishes, helmets and sports accessories, compact disks and bottles for milk.

There are various polymers (including copolymers) in this family but the most common polycarbonate is that based on bisphenol-A, also known by the commercial names — Lexan (U.S.) or Macrolon (Germany). Several polyblends and alloys are already in use, combining PC with ABS, PET, PBT and others.

Polyphenylene-Oxide (PPO, Noryl)   PPO appeared in 1964 as a conjugated oxidation product of phenolic monomers.

$$-\underset{CH_3}{\overset{CH_3}{\underset{\diagdown}{\overset{\diagup}{\bigcirc}}}}-O-$$

This polymer has a high softening temperature (175° to 190°C) and excellent mechanical properties. Due to difficulties in processing, a modification with polystyrene was developed in 1966 as an amorphous polyblend, mostly known by its commercial name of Noryl (by GE). Its mass density is 1.06. Noryl[TM] is one of the leading engineering polymers. The mechanical properties are retained over a broad temperature range — stable dimensions, low water absorption, light weight, nonburning and it is easily processed. Noryl is cheaper than the original PPO, so that it competes successfully with the other engineering polymers.

It may be conveniently processed via extrusion, injection and blow molding, as well as structural foaming. It is customary to reinforce with up to 30% glass fibers, and it is utilized in machinery and pumps, dishes (that may be directly heated on the fire), and various elements in transportation. Novel developments have led to a higher heat deflection temperature of 150°C, and enhanced impact strength.

Polysulfones (PSU)   This novel engineering polymer was developed in 1965 and has the following structure, specific mass = 1.24:

$$-[O-\bigcirc-\underset{CH_3}{\overset{CH_3}{\underset{|}{\overset{|}{C}}}}-\bigcirc-O-\bigcirc-SO_2-\bigcirc-]_n$$

It is obtained through a condensation reaction

$$\text{Na}^- \bigcirc\!\!\!\bigcirc -\overset{\overset{\displaystyle CH_3}{|}}{\underset{\underset{\displaystyle CH_3}{|}}{C}}- \bigcirc\!\!\!\bigcirc -O-\text{Na} + \text{Cl}- \bigcirc\!\!\!\bigcirc -SO_2- \bigcirc\!\!\!\bigcirc -\text{Cl}$$

It represents a broad family, based on the occurrence of sulfonic groups (SO_2) directly bonded to aromatic rings. Another modification led to the production of polyether sulfone (PES) which contains no aliphatic groups and has some advantages like improved impact strength and softening temperature. These are very stable amorphous polymers (due to their unique structures) that may be processed at elevated temperatures (315°C to 400°C) by extrusion, injection and blow molding, as well as thermal forming. Both mechanical and thermal properties are superb (heat-deflection temperature of 175°C), in addition to high transparency and self-extinguishing properties. It competes mainly with polycarbonate, although it is much more expensive. It is also distinguished by high chemical resistance even at elevated temperatures, very low creep, high dimensional stability and good electrical properties. On the other hand, the polysulfones do suffer from low resistance to UV radiation. Hence, it is suitable for very specific applications, when this rare combination of properties is required in electronics, household items, machinery and transportation. In practice the polysulfones have penetrated into a domain which was traditionally occupied by ceramics, glass or metals. Among the novel polysulfones also appears polyaryl-sulfone.

PES was introduced in 1972 by ICI, and its utilization has increased steadily. The chemical structure is

$$- \bigcirc\!\!\!\bigcirc -\overset{\overset{\displaystyle O}{\|}}{\underset{\underset{\displaystyle O}{\|}}{S}}- \bigcirc\!\!\!\bigcirc -O-$$

PES has T_g = 230°C as compared to 185°C for PSU and 295°C for polyaryl-sulfone.

Phenoxy Phenoxy is a thermoplastic polymer obtained by reacting bis-phenol-A with epichlorohydrin, reaching a molecular weight of around 25,000 ($\overline{M}_n$), and specific mass of 1.18.

$$-[O- \bigcirc\!\!\!\bigcirc -\overset{\overset{\displaystyle CH_3}{|}}{\underset{\underset{\displaystyle CH_3}{|}}{C}}- \bigcirc\!\!\!\bigcirc -O-CH_2-\underset{\underset{\displaystyle OH}{|}}{CH}-CH_2]_n- \qquad n = 10$$

Actually, the same reaction leads (at different conditions) to the thermosetting polymer epoxy, in which the ring (which is cross-linkable) is preserved.

$$-HC\underset{\diagdown\ \diagup}{\overset{O}{\frown}}CH-$$

Phenoxy is amorphous and transparent. It resembles, both in structure and properties, polycarbonate, except for the existence of the functional and reactive radical, OH, which may also react with di-isocyanate. Phenoxy excels in low permeability to oxygen and dimensional stability, however, it softens at 85°C. It may also crack under load, and absorb water. On the other hand, it is rigid and tough, easily processed via extrusion, injection, and blow molding at moderate temperature (200°C–270°C). Phenoxy is used in coatings and glues, as well as in some engineering applications, in spite of competition with polycarbonate.

Polyphenylene-Sulfide (PPS)   Developed in 1968, PPS was commercially introduced in 1973 (Ryton). This is a prominent engineering polymer, of the HT family, withstanding very high temperatures (around 280°C). It is considered borderline between thermoplastic and thermosetting. Its basic structure consists of an aromatic core bonded to sulfur in the para-position.

$$[-\hexagon -S-]_n$$

It is produced by reacting para-dichlorobenzene with sodium sulfide. This structure enables high crystallinity, dimensional stability, chemical and environmental resistance, and it is self-extinguishing. It is insoluble below 200°C. It may be processed mainly by injection at high temperature, while the mold itself must be kept at relatively high temperature. PPS has special applications in electronics and machinery when the unique combination of chemical inertness and excellent thermal and mechanical properties is desired. It is recommended for fiber reinforcement and is, therefore, usually supplied as a composite material. It is also used for special coatings. The price of PPS has dropped in line with growth in production, representing the most useful polymer among the premium engineering plastics.

Polyimides (PI), Polyamide-Imide (PAI), Polyether-Imide (PEI)   The typical imide group consists of

$$-N\underset{\diagdown}{\overset{\diagup}{}}\begin{matrix}CO-\\ \\CO-\end{matrix}$$

It is a condensation product between an aromatic diamine and an aromatic dianhydride, first developed in 1964 (Kapton).

$$- \text{⬡} - \text{O} - \text{⬡} - \text{N} \underset{\text{CO}}{\overset{\text{CO}}{<}} \text{⬡} \underset{\text{CO}}{\overset{\text{CO}}{>}} \text{N} -$$

There are several other related structures, some of which are thermosets. The significance of these polymers is their physical stability at peak temperatures of 300°C and above, together with high chemical stability. There are combinations of polyimides with polyamides and other modifications that aim at improved processability. The original polyimide is very difficult to process, being shaped only via compression or film casting, or by fiber spinning from solution.

Polyimide is one of the most stable polymers appearing reinforced (often by graphite) for specific engineering uses, or as a reinforcing fiber. It also appears as a plastic in electrical and engineering applications, but its cost is very high. By eliminating hydrogen, special polymers were developed which soften at an unbelievable temperature of 590°C. Currently, thermosetting polyimides are increasingly used as premium engineering materials. In order to improve the workability of PI, new variations in this family have been developed, gaining much importance in the domain of high performance engineering polymers. Most prominent are polyamide-imide (PAI) and polyether-imide (PEI). The structure of PAI is as follows:

$$- [\text{OC} - \text{⬡} \underset{\text{CO}}{\overset{\text{CO}}{<>}} \text{N} - \text{R} - \text{NH}]_n$$

It is a condensation product of trimellitic anhydride (TMI) and a mixture of diamines. PAI shows T_g of 275°C, and its superb mechanical properties are retained at high temperature, 260°C. It was introduced in 1976, mainly for injection molding. In order to further improve workability, the polymerization is stopped at low conversions, wherein the product is shaped by injection or extrusion and further hardened through continuing polymerization in an oven at a high temperature (postcuring). This process also leads to partial cross-linking, which is, however, undesirable for recycling.

Polyether-imide (PEI) combines structural stiffness (aromatic imide) together with easy flow and workability, due to the existence of the etheric bonds. It appeared initially in 1982 under the commercial name Ultem (GE).

$$- [\text{N} \underset{\text{CO}}{\overset{\text{CO}}{<>}} \text{⬡} - \text{O} - \text{⬡} - \underset{\text{CH}_3}{\overset{\text{CH}_3}{\underset{|}{\overset{|}{\text{C}}}}} - \text{⬡} - \text{O} - \text{⬡} \underset{\text{CO}}{\overset{\text{CO}}{<>}} \text{N} - \text{⬡}]_n -$$

The polymerization is carried out in two stages—initially the aromatic polyimide is obtained and in the latter stage, etheric bonds are introduced. The polymer is thermoplastic. PEI represents an unique engineering polymer exhibiting a superb combination of mechanical and thermal performance, easy processibility together with a reasonable price. It also withstands UV radiation, fire and many chemicals, but may be sensitive to alkalinity. Mainly reinforced, it serves as a substitute for metals in first class electronic and engineering appliances.

Ionomer   This unique polymer was introduced in 1964, under the commercial name Surlyn. It has an unusual chemical structure, rendering it as a hybrid between thermoplastics and thermosets. Ionomer is a copolymer of ethylene and methacrylic acid combined with metallic (monovalent or divalent) ions.

$$-(CH_2CH_2)_m-(CH_2\underset{\underset{COO^-M^+}{|}}{\overset{\overset{CH_3}{|}}{C}})_n- \qquad M^+ = \text{metallic ion}$$

The ionic bond with the metal induces temporary cross-linking at ambient temperatures, while upon heating the ionomer behaves like other thermoplastic materials. This behavior is very desirable—the easy workability of thermoplastics together with the dimensional stability of thermosets—a reversible transition. Ionomer is a tough transparent polymer, appearing in various degrees of crystallinity or ionization. Principally it resembles polyethylene in mechanical properties (specific mass of 0.94), but surpasses it in stability, resistance to ESC and transparency. Like polyethylene it is easily processed via all common methods. It appears often in multilayered packaging, containers and coatings.

Poly-Benzimidazole

Like polyimide, poly-benzimidazole is obtained upon reacting diamines, but with different proportions. It was developed in 1964 as a premium and unique HT engineering polymer, stable at 300°C and calling for special processing methods such as plastomer or fiber.

Poly-Para-Xylene (Parylene)

$$-H_2C- \langle\bigcirc\rangle -CH_2-$$

This polymer is obtained by an addition reaction of para-xylene, via pyrolysis at extremely high temperatures. It was developed in 1965 and various modifications have appeared since then. It is an engineering polymer for specific applications, withstanding elevated temperatures (softening between 260°C and 400°C).

Polyarylate (PAR)   Polyarylate is a premium engineering polymer, initiated in 1974 as a polyester obtained by reacting diphenol (bis-phenol A) with an aromatic di-acid, (a mixture of iso- and tera-phthalic-acids). It is commonly known by its commercial name of Ardel (Union Carbide™) and is an amorphous polymer with excellent mechanical, optical and thermal properties. It excels also in weather resistance, so it can serve as the exterior layer on top of engineering polymers. In contrast to PC, it is not susceptible to loss of high impact strength as result of residual stresses. It is, therefore, not so sensitive to the thickness of the product, the notch radius or annealing conditions. It also has high resilience, so that it is becoming a central member of the high-performance engineering polymers.

Polyether-Etherketone (PEEK)

$$- \langle\bigcirc\rangle -CO - \langle\bigcirc\rangle - O - \langle\bigcirc\rangle - O-$$

This semi-crystalline polymer belongs to the family of polyether ketones, $T_g = 145°C$; $T_m = 235°C$. Developed in 1980 by ICI, it is a superb engineering polymer, showing excellent mechanical properties that are retained at elevated temperatures. Due to its extremely high price, its utility is still limited to the field of aviation and space (reinforced with carbon fibers), electronics and machinery. Another advantage is its stability towards fire or chemicals, although it is sensitive to UV radiation.

Liquid Crystal Polymers (LCP)   This novel polymeric family excels in thermal and mechanical performance. The uniqueness of these polymers stems from the extraordinary crystalline structure, exhibiting ordered domains even in the liquid state. These are chainlike macromolecules of rigid structure, mainly because of the existence of aromatic rings such as aromatic polyamides and polyesters. There are two major groups — lyotropic and thermotropic. In the former group, the liquid crystals are formed in an appropriate solution, usually forming fibers (like Kevlar™, that appeared in 1965 as an aramide).

The thermotropic group consists of a crystalline liquid formed during cooling of a dry melt (similar to the processing of thermoplastics)—the first having been the aromatic polyesters (1972). When a disordered melt cools in a region of proper transition temperature (T_l), a liquid crystal phase (mesophase) is formed. It is characterized by high order (anisotropic) that creates turbidity. The most common form is the nematic, a bundle of parallel, long, rodlike molecules. Additional cooling to the primary transition temperature, T_m, leads to solidification into a solid crystalline phase (small crystallites). In the region between T_l and T_m, a liquid of very low viscosity prevails, in contrast to the high melt viscosity. The aromatic polyesters have high heat distortion temperatures (HDT). Vectra, composed of para-hydroxy-benzoic acid (PHBA) and para-hydroxy-naphtoic acid (PHNA), has an HDT of 180°C–240°C. Xydar, composed of PHBA, tera-phthalic acid and biphenol, has an even higher HDT of 260°C–350°C.

Cellulosic Polymers   This group of polymers is at the borderline between natural and artificial—better defined as modified natural polymers. The raw material for these polymers is cellulose ($C_6H_{12}O_4)_n$ abundantly found in nature as wood, straw or cotton. By chemical treatment various derivatives may be obtained, which perform like thermoplastics. The first one in this group, that also happens to be the oldest polymer (created in 1870), is cellulose nitrate. This is a hard and transparent material, sensitive to heat and solvents (also to fire) and unable to be processed by the regular processing methods for thermoplastics (extrusion and injection). Therefore, it was plasticized by camphor, thus producing celluloid, which is found in commodities like drawing sets, pens and pencils, buttons and also in coatings and glues. However, its use is gradually diminishing.

Among other derivatives of cellulose are cellulose-acetate (CA) which appeared in 1927, ethyl-cellulose (1935), cellulose–acetate–butyrate (CAB) developed in 1938, and cellulose-acetate-propionate (CAP) developed in 1960. The acetate exceeds the nitrate in stability and resistance, and serves as photographic film, scotch tape, frames for glasses, handles and packaging. A distinct disadvantage lies in its high water absorbancy. CAP and CAB exhibit good workability and weatherability.

Ethyl cellulose exhibits the best properties within the family and is highly exploited. In conclusion, the cellulose family played a significant role at the beginning of the polymer era, but now the younger polymers with improved properties and workability, as well as availability of raw materials for mass production, compete successfully.

Thermosets

Thermosets represent about 15 to 20% of polymer production, however, they have unique properties and performance.

Phenol-Formaldehyde (PF)   This is the oldest completely synthetic polymer (made in 1909), and still appears as a useful and cheap material. It is also known by its initial commercial name Bakelite™ (named after its inventor Bakeland). PF is a condensation product between phenol

⬡ $-OH$ and formaldehyde $CH_2{=}O$

After cross-linking it reaches a three-dimensional structure:

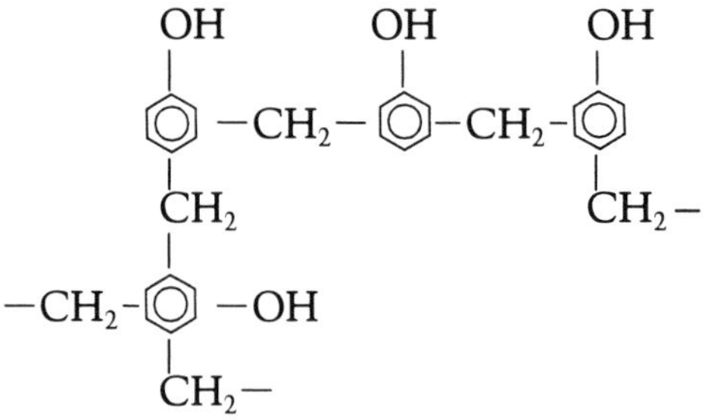

Different proportions of the two reactants produce different intermediate products, which behave like thermoplastics and are therefore processed and shaped via compression (as a fluid) before achieving the ultimate thermosetting state. When an excess of phenol is used (1 mol of phenol to 0.8 mol of formaldehyde) and an acidic catalyst, Novalak™ is obtained. Ortho and para bonds, formed at a molecular weight of 1200–1500, are eventually cross-linked by adding hexamethylene-tetramine (or paraformaldehyde) and catalysts.

Intermediate stages:

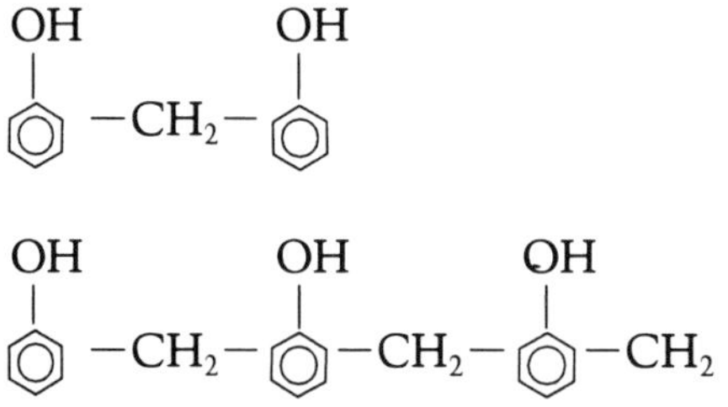

Another form, named Resol™, is based on excess formaldehyde (1.25 : 1) and a basic catalyst. Resol contains many functional groups CH_2OH which are cross-linked at elevated temperatures. Hence, Novolak is cured in two stages while Resol™ needs only one stage.

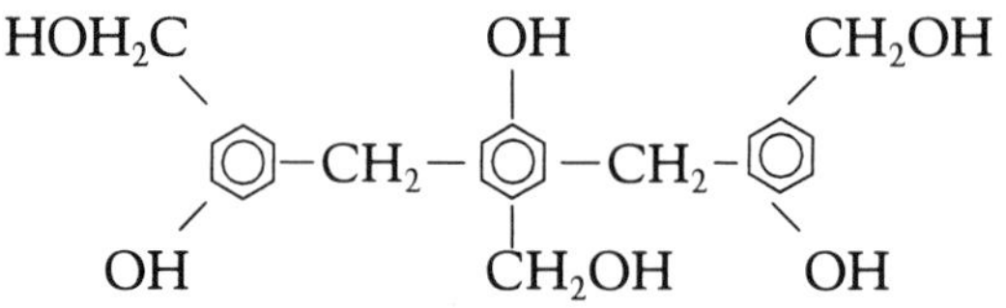

Resol™   MW = 300–700

When the molding powder is introduced into a mold and high pressure and temperature are applied (up to 100 atmospheres, 160°C–200°C), the melt fills the mold and then hardens to a thermosetting state. Curing time lasts usually several minutes. Here, fillers are essential, without which a brittle plastomer is obtained. Fibrous fillers are commonly used such as wood dust, cotton, asbestos or glass fiber. The type of reinforcement dictates the price as well as the mechanical, thermal and electrical properties.

Reinforced phenol-formaldehyde excels in impact strength, hardness, rigidity, chemical or thermal stability and high dielectric constant. Asbestos fibers improve thermal stability up to 200°C, as compared to 115°C with wood dust. On the other hand, asbestos also increases heat conduction. The use of asbestos is restricted however, due to suspicion of carcinogenic properties. Improved mechanical and thermal properties are achieved by reinforcement with glass fibers. The reinforced plastic is relatively heavy (specific mass 1.4 to 1.9). Phenol-formaldehyde is utilized solely for dark items as it darkens with time (and exudes phenol). The weatherability is fair. This polymer serves as a material of construction in machinery and electrical equipment, as well as glues and coatings. The molding powder is cast or compressed to many electrical appliances, ashtrays, radio cabinets, handles and tools, furniture, accessories in laundry machines and telephones. The phenolic resin appears also as a binder in plywood. It is much used in the production of laminates that consist of layers of paper or fabric soaked with the resin, pressed under high pressure. These laminated sheets are very useful for construction or ornamentals and can be shaped by milling.

Foamed phenolic polymer appears to be an excellent insulator, stable and cheap. There is much competition for similar outlets between thermosets and thermoplastics—urea and melamine, ABS, acetal, PP and Nylon. The phenolics have the advantage of low price, and a good combination of general performance when appropriately composed. On the other hand, use of thermoplastics is frequently preferable due to improved processing in mass production, in addition to the recyclability of wastes.

Urea-Formaldehyde (UF, 1928), Melamine-Formaldehyde (MF, 1939)   UF is the condensation product of the reaction between urea and formaldehyde.

$$O{=}C{-}NHC\,H_2{-}O{-}CH_2HN$$
$$\diagdown$$
$$C{=}O$$
$$\diagup$$
$$NH{-}H_2C{-}HN$$
$$\diagup$$
$$O{=}C$$
$$\diagdown$$
$$NH{-}CH_2{-}O{-}CH_2{-}NH$$
$$\diagdown$$
$$C{=}O$$
$$\diagup$$
$${-}O$$

MF is a condensation product between melamine and formaldehyde.

$$
\begin{array}{c}
\text{NH} - \\
| \\
\text{C} \\
\diagup\diagup \quad \diagdown \\
\text{N} \quad \text{N} \\
| \quad\quad || \\
\text{C} \quad \text{C} \\
\diagup \quad \diagdown\diagup \quad \diagdown \\
-\text{HN} \quad \text{N} \quad\quad \text{NH}-\text{CH}_2-\text{CH}_2\text{NH}-
\end{array}
$$

There is a marked similarity in synthesis and utility between urea and melamine formaldehyde. Both amino resins are thermosets obtained via condensation reactions between amine (NH_2) derivatives and formaldehyde. UF is the older polymer. In contrast to PF, this polymer is bright and translucent. It surpasses PF in other properties, but is also more costly. Reinforcement is ordinarily achieved with treated cellulose. In addition to use in lighting and electrical appliances, in radios and in the household, UF resins are frequently used in protective coatings (as for cars) or premium thermosetting glues. Foamed UF is also often used for thermal insulation. MF, as compared to UF, excels in water and heat resistance and it has penetrated primarily into the field of laminates. The upper layers of Formica™ are composed of paper soaked with MF, on top of other layers soaked with PF. The melamine laminates exhibit high clarity, gloss, hardness, mechanical and thermal strength. They meet the challenge of the market for table dishes, exhibiting clarity, gloss, strength and stability. MF when reinforced with various fillers is much used for appliances, and, like UF, is also widely used in glues and coatings.

Unsaturated Polyester (UPES)   In 1942, unsaturated polyesters were introduced. They result from condensation reactions between diols, (like ethylene glycol HO$-$CH$_2-$CH$_2-$OH or propylene glycol), and a blend of saturated and unsaturated di-acids, (like phthalic acid or anhydride, maleic and fumaric acids and others).

$$
\begin{array}{cc}
\begin{array}{c}
\text{O} \\
||\\
\text{C} \\
\diagup \quad \diagdown \\
\bigcirc \quad\quad \text{O} \\
\diagdown \quad \diagup \\
\text{C} \\
||\\
\text{O}
\end{array}
&
\begin{array}{c}
\text{O} \\
||\\
\text{C}-\text{CH} \\
\diagup \quad\quad || \\
\text{O} \quad\quad\quad || \\
\diagdown \quad\quad\quad \\
\text{C}- \text{CH} \\
||\\
\text{O}
\end{array}
\end{array}
$$

(phthalic anhydride)   (maleic anhydride)

If only maleic acid is used, a linear chain is reached in the first step,

$$
\text{HO} -(\text{CO}-\text{CH}=\text{CH}- \text{CO}-\text{O}-\text{CH}_2\,\text{CH}_2\text{O})_n\text{H}
$$

The chain may eventually cross-link through an addition polymerization with another monomer (usually styrene via the double bond). There are various polyesters for different uses, however, fiber reinforcement (mostly glass fibers) must be applied in any case. Reinforced polyester is very popular as a construction material for boats and cars, as well as in the aviation industry. Glass fibers (in various shapes and grades) contribute extraordinary mechanical strength (up to 40% by weight of the composite). The polyester resin serves as the binder that bonds and unifies the glass fibers, thus enabling the final shaping, mostly by casting. The resin appears as a liquid and is amenable to be cast or sprayed on top of the glass fibers. Care must be taken to provide good adhesion between resin and fibers. Frequently, a special binder is applied. It is essential to cover the fibers, protecting them against humidity which may weaken the mechanical strength. The polymer is cured with the aid of catalysts and heat. Appropriate catalysts may also be utilized at ambient temperature. The reinforced polyester is a strong and stable material with good optical properties (translucent to opaque). It has excellent mechanical and thermal properties, as well as good electrical and environmental properties (provided UVA and flame retardants are applied). This is the most promising polymer in construction. Huge structures are mostly handmade, but sheets have already been made on continuous moving belts sometimes involving heat curing. It is widely used in construction (partitions, roofs, shutters, screens and furniture) and in industry (piping and tanks for fuel and chemicals, helmets). It has much utility in transportation—automobiles, boats and planes. Another common polyester is manufactured from saturated acids, glycol and glycerol, and modified by fatty acids, leading to alkyds (developed in 1926). Alkyds are currently the major component in synthetic coatings. Alkyds may also serve as a thermosetting plastomer.

Epoxy Resins (EP) Epoxy was developed in 1947 as a premium thermoset and is based on a condensation reaction between bis-phenol-A and epichlorohydrin, characterized by the cyclic ether group

$$-HC\underset{\diagdown\diagup}{\overset{O}{\frown}}CH-$$

$$CH_2\overset{O}{-}CH-CH_2-[O-\bigcirc-\underset{CH_3}{\overset{CH_3}{\underset{|}{\overset{|}{C}}}}-\bigcirc-O-CH_2-\overset{OH}{\underset{|}{CH}}-CH_2]_n-O-$$

$$\bigcirc-\underset{CH_3}{\overset{CH_3}{\underset{|}{\overset{|}{C}}}}-\bigcirc-O-CH_2CH\underset{\diagdown}{\overset{O}{\frown}}CH_2$$

The prepolymer has n = 0–14 (the lower value is a liquid and the upper a solid), wherein the maximum MW is 4,000. These turn into thermosets after adding hardeners like polyamines. Epoxy appears in several ranges of molecular weight, from liquids (suitable for glues and castings) to solid resins (useful as solutions in protective coatings). The basic polymer may also be modified. Various hardeners are in use both at high temperature or room temperature. Major properties depend on the type of epoxy and the reinforcement. Epoxy exhibits excellent bonding properties, and is useful for protective coatings and reinforced construction materials. The most important properties include high chemical resistance (mainly to alkali), high tensile strength, and durability at elevated temperatures, in addition to dimensional stability and good electrical properties.

In the domain of glues (epoxy is also known by the commercial name Araldite^{TM}), there is no substitute for the bonding of metals (like aluminum). An appropriate composition provides toughness and flexibility, as in a blend with thiokol rubbers or polyamides—although there is diminished thermal and chemical performance. Protective coatings made from epoxy withstand relatively high temperatures, solvents and corrosive chemicals. The forming of plastomer is frequently achieved by casting. Reinforced epoxy (usually with glass fibers) surpasses the properties of reinforced polyesters, but is more costly. It is used in aviation, tanks and tubes, accessories, and even molds. It also serves as a high-quality electrical insulator. The primary modern outlet is the building industry.

Polyurethane (PUR)   The characteristic group is

$$
\begin{array}{cc}
H & O \\
| & \| \\
-N-C-O-
\end{array}
$$

It is obtained through a reaction between di-isocyanates and glycols, that are supplied as polyesters or polyethers. It was developed by Bayer in Germany in 1937, and later in the United States (1953).

$$
n\,OCN-R-NCO + n\,HOR'\,OH \rightarrow (RNH-COO-R'-O\overset{\text{O}}{\overset{\|}{C}}NH)_n \quad (6\text{-}3)
$$

This is an addition reaction that proceeds via a mechanism of step growth (stagewise) polymerization. The most common isocyanates are toluene diisocyanate (TDI) and methylene bis 4-phenylisocyanate (MDI). TDI is cheaper and is mostly used for flexible foams, but MDI competes because of its lower vapor pressure and reduced sensitivity in handling. Polyfunctional isocyanates are added for cross-linking. The interesting reaction between isocyanate and water evolves CO_2, which is the preferred method for obtaining flexible foams.

$$R-NCO + H_2O \rightarrow R-\overset{H}{\underset{|}{N}}-\overset{O}{\underset{\|}{C}}-OH \rightarrow RNH_2 + CO_2 \qquad (6-4)$$

Isocyanates also react with amines:

$$RNH_2 + R'-NCO \rightarrow R-\overset{H}{\underset{|}{N}}-\overset{O}{\underset{\|}{C}}-\overset{H}{\underset{|}{N}}-R' \qquad (6-5)$$

In order to obtain a rigid foam, a polyester with a short but highly reactive chain, MDI, and a physical foaming agent (a volatile liquid like Freon) are mixed. For a flexible foam, a long-chained polyether and TDI are mostly used. The most important use of PUR is as a foam, either flexible or rigid. In addition it is found in coatings, glues, elastomers or rigid plastomers. Urethane paints are offered for special anticorrosive uses. Polyurethane elastometers have penetrated the market well, in shoes and also in transportation (tires). A soft polyurethane foam is obtained during polymerization when water is added to release the gas carbon dioxide from the isocyanate. By appropriate control of the combination of reactants, products are reached exhibiting various densities (in the range of 16 to 45 kg per cubic meter) and a structure of open pores. This type of foaming does not need heat, pressure or external foaming agents. This is the advantage of polyurethane over other foamed plastomers. The mechanical and thermal (insulation) properties depend on the density of the foam. Flexible foamed polyurethane is strong and durable and is used often in upholstery, cushions, mattresses, in transportation, or in homes. Rigid polyurethane foam serves in transportation as well as in insulating refrigerators or cold-storage (at a density range of 130–160 kg per cubic meter with an open-cell structure). Semi-rigid or structural (integral) foams are also common.

Silicones (SI)   This is a group of materials differing in structure from the rest of organic compounds, due to the presence of the inorganic group Si-O-Si, based on silicon atoms Si which replace carbon in the main chain. The side groups, however, are paraffinic.

$$-\left[\underset{\underset{R}{|}}{\overset{\overset{R}{|}}{Si}}-O\right]_n-$$

where $R = CH_3; C_2H_5$.

Silicones appear in various compositions from thin fluids up to polymers that act as elastomers or thermosetting rigid plastomers. Silicones excel in weather durability, water repellency and in electrical and thermal properties. Silicone resins or coatings have heat-distortion temperatures up to 450°C. They are much used in aviation and the space industry as electrical insulation for motors, glue, protective coatings, a premium elastomer or as additives in construction.

Allyl, Poly-Diallyl-Phthalate (PDAP)   This was developed in 1941 as an unsaturated polyester obtained from diallyl phthalate,

$$CH_2 = CH-CH_2-OH \quad \text{(allyl alcohol)}$$

This is a thermosetting polymer, resembling phenols in its properties, but exhibiting high resistance to alkali and solvents, together with a retention of electrical properties under extreme conditions of heat and humidity. It is mostly used for electronics, coatings and laminates.

Furan   This is a condensation product involving furfuryl alcohol which may compete with phenol formaldehyde in industrial processes. The advantage stems from the raw material furfural that originates in plants such as straw, corn, sugar cane or wood. It is not however much in use in developed countries. The polymer itself shows good chemical resistance and serves in special coatings, adhesives and laminates. It is supplied as a relatively cheap fluid resin, hardened by the use of an acidic catalyst.

6.2 SPECIAL USES

After describing in detail some 40 families of polymers, it is now appropriate to sort them according to their uses. Five thermoplastic polymers lead in regard to use and relative low cost—polyethylenes (HDPE or LDPE and LLDPE), PVC, polystyrene and polypropylene—amounting together to around 70% of total consumption. PET is also sometimes considered as a commodity (for bottles). A second group consists of more expensive polymers used for specialties—ABS, acrylics, cellulose or several thermosets. A unique group is represented by the engineering polymers, most prominent being polyacetal, Nylon™, polycarbonate, Noryl, linear polyester (mainly PBT), polysulfone, polyphenylene-sulfide, polyimide and fluorocarbons. These are usually distinguished but rather expensive polymers.

The novel polymers (end of 1960s and early 1970s) belong mostly to the HT group and their overall utility is limited. Many appear as special fibers, like aromatic polyamides and polyesters, polyimides and Parylene. The novel structures have rings (either aromatic or heterocyclic), in the main chain (para-phenylene is preferred over meta-phenylene). Substituting C—H bonds with C$-$N, C$-$O, C$-$S bonds or ladderlike structure (needing the breakdown of two adjacent bonds for degradation) is typical in these structures. By pyrolysis of some polymers (polyacrylonitrile or polyamide), carbon fibers may be obtained. Most recent polymers include PEEK (and its derivatives, polyether ketones) and the liquid crystal polymers (LCP). A common division of polymers is as plastomers (plastics), elastomers, fibers, coatings and adhesives. Therefore, these groups will be briefly described with most attention devoted to plastics.

Synthetic Elastomers

The synthetic elastomers (due to improved performance) are considered as first-class rubber materials on their own merit, and not just as a substitute for natural rubber. Consequently, polymers with quite a different composition from natural rubber are considered as elastomers. What determines their utility is mostly their rubberlike properties—resilience that is expressed as an extensive elongation in tension (500% to 1000%) quickly recoverable (almost to the original dimensions) when stretching is stopped. Other essential properties consist of resistance to abrasion, high tear strength, impact strength and resistance to fatigue. It is important to note thermal stability, as well as resistance to oils, solvents and the environment. The strength of elastomers may be augmented during stretching by orientation and crystallization. The synthetic elastomers excel in toughness and retraction. They are long chains in an amorphous phase. They are basically thermoplastic but undergo slight cross-linking; thus binding the chains together in order to eliminate slippage during stretching. A typical analogy is a natural rubber that needs slight sulfurization (namely, partial cross-linking with sulfur), in order to achieve the desired mechanical properties. A high degree of cross-linking leads to a rigid rubber (ebonite) which is actually considered a thermoset plastomer.

The elastomer must exhibit a low value of T_g. Among polymers that might be regarded as engineering elastomers the following should be mentioned—butadiene-styrene copolymer (GR-S or SBR), butyl, neoprene, EPR (copolymer ethylene-propylene), nitrile, polybutadiene, thiokol, polyisoprene, silicon, polyurethane, Hypalon, and EPDM. The internal breakdown by consumption is about 75% synthetic versus 25% natural rubber. Within the family of synthetic elastomers a typical breakdown is about 46% SBR, 19% polybutadiene, 9% EPR, 4% neoprene and 3% nitrile.

Natural rubber itself is a polyisoprene (cis). By the addition of carbon black, it reaches excellent mechanical properties; however, natural rubber suffers from relatively low resistance to heat, oxygen from the atmosphere (mainly to ozone), and various oils or solvents. The synthetic polyisoprene and SBR are typical rubber substitutes, resembling rubber in general performance.

The primary need for a substitute arose in the Second World War, when the supply of rubber from the natural sources was abruptly cut down. SBR slightly surpasses natural rubber in abrasion resistance and external endurance, but it builds up more heat as a result of the absorption of heat (hysteresis). SBR also needs fillers like carbon black and cross-linking agents. Neoprene exhibits excellent resistance to oils, solvents and heat—it is non-burning and weather resistant. Thiokol also withstands oils and solvents, but is inferior in mechanical properties. Hypalon™ is obtained by reacting chlorine and sulfur dioxide with polyethylene and the resultant elastomers excel in mechanical properties and weatherability. The silicones withstand high temperatures and chemicals, but their mechanical properties are only moderate. The elastomeric polyurethanes combine successfully the properties of rigidity and good ductility and they too are weather resistant. Butyls show low gas permeability and are therefore highly recommended for inner tubes of tires. The ethylene-propylene (EPR) copolymer is a relatively modern and promising synthetic rubber, because of its reasonable general performance at a low price. The engineering uses of elastomers are in transportation (mainly tires), piping, conveying belts, packaging, sealing, electrical insulation and many appliances. They are also used for internal coating in pumps, valves and in containers for storage or transporting of fuel and chemicals.

EPR is currently replaced by EPDM, a modification with a diene monomer, due to its improved workability. A novel type of elastomer (called a thermoplastic elastomer) exhibits quite revolutionary behavior. Here cross-linking is temporary (at room temperature) while it can flow at higher temperatures, like thermoplastics. The typical one (SBS) is a strictly ordered block copolymer of styrene and butadiene, made by an anionic polymerization. The butadiene chains (at a controlled MW of 70,000) are embedded in a rigid phase of polystyrene spheres (MW of 15,000) thus providing temporary cross-linking at ambient conditions, while being processible at high temperatures like thermoplastics.

Synthetic Fibers

Various synthetic fibers appear in clothing, upholstery, and industrial uses. They are better known by commercial names, that hide their source and composition. Quite often a blend of natural and synthetic fibers is offered. The first man-made fibers (that still are of major use) are essentially based on a modification of natural cellulose. Most common in use are rayon (viscose) and cellulose-acetate (called acetate). The oldest synthetic polymer in the textile industry is the polyamide (Nylon 6-6) developed in 1935. Currently there are many synthetic fibers, like the following:

Saran (copolymer of vinylchloride and vinylidene-chloride).
Orlon (polyacrylonitrile).
Acrylan (copolymer acrylonitrile).
Dynal (copolymer of acrylonitrile and vinylchloride).
Dacron or Terylene (a linear polyester based on teraphthalic acid).

Among other polymers that appear also as fibers are PE, PP and Teflon. The synthetic fibers have high strength in the axial direction, reached through cold stretching of the crystalline polymer. The strand itself is prepared by weaving long fibers, produced by extruding the polymer as a melt or solution.

The requirements from a premium fiber for clothing are high flexibility, resistance to tear, endurance at high temperatures (ironing), low water absorbency and low gloss. It is also essential to acquire easy weaving, conditioning and dying properties as well as insulation, permeability, weatherability, a pleasant touch to the human body and resistance to moths and perspiration. Other important characteristics are low contraction during laundering, low wrinkling (wash and wear), easy laundering and extensive resistance to detergents. Nylon is acknowledged as a fiber of high strength, flexible and stable, enduring relatively high temperatures, and is resistant to moths and perspiration. On the other hand, its weatherability is rather unsatisfactory. It has wide and versatile utility in industry for threads and ropes (internal-layer in tires, threads for fishing rods) cables and nets and in clothing (mainly, stockings, underwear and shirts).

Dacron (Terylin) surpasses Nylon™ in clothing, due to its resistance to chemicals. It does not contract or stretch during laundering. It is much used for dresses, suits and curtains. Orlon has a lower abrasion resistance, however, it excels in insulation and resistance to light and solvents. One should remember that most synthetic fibers may be selectively sensitive to solvents which should be considered during dry cleaning. Orlon™ does not wrinkle and is used for blankets, sweaters, rugs and general textiles.

Saran is stable in water and withstands high temperatures. It is supplied as films or woven and it is extensively used for upholstery in household or transportation. It is found in interior clothing, suits, rugs, and tire reinforcement. Rayon suffers from high water absorbency and is frequently used as a blend with other fibers. An interesting product is found in PET (recovered from bottles) as a textile for interior and insulating clothing. Acetate is found as a blend in fine clothing and upholstery. Artificial fibers like Nylon and others can be used as a reinforcement to plastomers. A typical breakdown is shown in Table 6-2.

The novel fibers serve as reinforcement for special performances like endurance at high temperatures and premium mechanical properties. They consist of aromatic polyesters and polyamides, polyimides and other high-performance polymers. Most distinguished are Kevlar™, Nomex™, Ekonol™ and PEEK™.

Synthetic Coatings and Paints

Most polymers may be dispersed as a solution or an emulsion, thus acting as a base for synthetic varnishes. In addition to the primary polymeric resin (the binder) and the dispersing fluid (organic solvents or water), paint consists also of pigments, fillers, plasticizers, catalysts and stabilizers. Domestic

TABLE 6-2
Consumption of Synthetic Fibers

	Percent use
Polyester	46
Nylon	25
Olefins (PE, PP)	9
Acrylics	6
Rayon	3.5
Acetate	1.6
Glass fiber	9

paints were historically manufactured by using drying oils (linseed or ricinoleic oils), sometimes slightly polymerized (thickened). In parallel many natural resins were used. Currently there is much use of alkyds, polyesters bound to fatty acids. Alkyds, bonded in various proportions to drying and semi-drying oils, are the leading component in domestic or industrial paints. On the other hand, the use of polyvinylacetate, acrylates and styrene-butadiene copolymer (all in aqueous emulsions) has become very popular for painting walls. Industrial and automotive paints are based on polyesters, phenol, urea and melamine formaldehydes, epoxides, polyurethane, vinyl, acryl, polystyrene, cellulose and silicones. From a practical point of view, there is a division between domestic paints (for protection or ornamentation on wood, concrete and metal), and industrial paints (for anti-corrosion purposes), and paints for transportation (cars, boats and planes), as well as varnishes for electrical insulation. Other uses consist of coatings on fabric, leather and paper.

The specific properties of polymers for paints are:
Easy solvation in relatively cheap solvents, or dispersion in water.
Easy dying with pigments.
Easy applicability by brush, roller or spray gun.
Quick drying and hardening as a thin film, which may either be a thermosetting gel or a thermoplastic.
Creation of a strong but flexible film, strongly adhering to the substrate, and resisting heat, chemicals or the environment.
Achievement of smooth and aesthetic films at various degrees of light reflection up to high gloss.

The paint itself must be stable during storage with no sedimentation or curing prior to use. There is a distinction between paints that dry at room temperature (mostly domestic and some industrial paints) and thermosetting paints that must be baked with heat. The latter appear in car or appliance painting, with urea or melamine polymers and others. Epoxy also cures at high temperatures, but special catalysts may be used for curing at ambient temperature. Epoxy is considered an excellent industrial paint, appearing in various compositions. It has excellent adherence and resistance to solvents and chemicals as well as the environment. The accelerator solution must be supplied in a separate package and added prior to use.

Copolymers of vinylchloride are much used for industrial paints as coatings for containers that withstand water and chemicals. Polyurethanes are relatively new as coatings, mostly for industrial purposes. Silicones are used for specific conditions at ultra high temperatures providing anti-corrosion protection and water repellence—their weather resistance is superb. Car painting is an important issue where the use of nitrocellulose is diminishing in favor of urea, melamine or acrylics. Polyesters provide cheap and strong coatings which may also be supplied solvent-free. The use of aqueous emulsions has many advantages—quick drying, cheap dispersers—while organic solvents (which may add health and safety hazards) are eliminated. Therefore, as a result of ecological motivation there is a shift towards aqueous and solvent-free paints. There are many special coatings, some blends of natural and synthetic resins. Bonding with the substrate and endurance under specified conditions dictates the application of the appropriate paint. Needless to say, treatment of the substrate or use of intermediate (base) coatings is essential.

Synthetic Adhesives

Adhesives are essential not only for bonding or fixing but also as the elementary component when designing and assembling construction units. They are widely used in building construction and aviation. The process of adhesion is cheaper than most other binding operations, like welding or nailing, and frequently surpasses the other means in mechanical strength of the bond. The required characteristics of polymers for adhesion are good wetting and penetration to the substrate, excellent adhesion, appropriate initial fluidity followed by rapid curing at ambient temperature (or by heat and pressure) with little contraction. After adhesion the glue should withstand water, chemicals and heat. Other requirements consist of bond flexibility, matching expansion coefficient and reasonable tensile and impact strength. Thermosetting adhesives offer the general advantage of improved resistance to heat and solvents. The oldest synthetic adhesives are cellulose and phenols. For wood and paper, the most popular adhesives are thermoplastic polyvinylacetate, PVA (cold glue), and the thermosets phenol, urea and melamine formaldehyde. In plywood, phenolic adhesives play a major role.

A most important industrial adhesive is epoxy. This resin acquires an unusual adhesion to all kinds of materials including metals (like aluminum) and glass. It is supplied as a two-package glue (resin and hardener). It is cured at elevated temperatures and, also, at room temperature. By special modifications, flexible adhesives are also obtained. Polyurethanes have also penetrated this field, while silicones provide adhesives for high temperatures. Various thermoplastics and elastomers, including cellulose, serve as cement. Adhesion is essential in laminates as well as in sandwiches of rigid and hollow bodies. Innovations in the domain of adhesives include melt adhesives, namely, solvent-free thermoplastic contact glues.

Uses in Construction   Construction is still the greatest challenge for plastics. Around 20% to 25% of all plastics are currently used as building elements, pipes, coatings, thermal, or electrical insulation and decoration. This application reached 5.4 million tons in the U.S. in the year 1990; however this amounted to less than 4% of the total building materials. Therefore, the use of polymers in construction is yet in its childhood and a significant expansion is anticipated. In spite of the relatively higher cost (mainly if compared to concrete), plastics provide a huge versatility in processing and coloring. In contrast to wood and common metals, appropriate plastics do not suffer from corrosion or humidity, and therefore need no special maintenance. Plastics, as a group, allow aesthetic, lightweight and easy workable complicated systems (such as sanitary and sewage design). On the other hand, plastics need weather and fire stabilization. By adding appropriate reinforcement, performance reaches that of light metals. Prefabricated construction elements which include a layer of foam, combine mechanical strength, rigidity, low weight and good thermal insulation.

Transparent polymers substitute for glass because of superb impact strength and resistance to thermal fluctuation. Due to high endurance, plastics are used in paints, coatings and piping. In addition, polymers appear as adhesives for wood and other materials of construction, in sealing roofs and covering of buildings. Some of the most useful plastics in construction are described here.

PVC is first, comprising 45% of all polymers. As a rigid polymer it serves in partitions, ceilings, shutters, doors and windows, sewage and waste piping, conduits for electrical wires and various appliances. Appropriate design needs to take into account the high coefficient of expansion and tendency for creep. The mechanical strength does not suffice for weight bearing walls, but suits partitions or wavy ceilings. Upon stabilization, weatherability is improved. Many colorful decorative options are possible in transparent, translucent or opaque partitions. In Europe (mainly Germany) use of PVC in window frames has appeared to be a great hit. Plasticized PVC serves in electrical insulation, wall coating and increasingly in flooring. It also appears in furniture upholstery.

Polyethylene reaches about 7% of consumption. In flexible form, (LDPE) mainly appears in electrical insulation, containers and as protective films against humidity. The rigid form (HDPE) is used in pipelines and accessories. Polyethylene films protect concrete against loss of water during curing as well as in greenhouses or storage in agriculture.

Formaldehyde derivatives include phenol formaldehyde as well as urea and melamine, and are used as a glue in plywood and laminates. Urea formaldehyde also appears as an insulating foam (fire resistant).

Reinforced (fiber glass) polyester is a common building material, that yields to PVC. It is useful in partitions, prefabricated sheets, pipelines, furniture and coatings. This unsaturated polyester is also an organic cement in flooring or adhesives.

Alkyds are mostly used in domestic paints for wood, metal and concrete.

Polystyrene is used as a thermal insulation (foam) and in wall paints.

Acrylics are found in sanitary systems (baths and toilets), as transparent ceilings, lamps and coatings.

Polycarbonate is used for glazing, windows and lighting elements.

Rigid polyurethane foam is mainly useful for thermal insulation.

Polyvinylacetate is used in wall paints (PVA emulsions).

Epoxy serves for concrete bonding, as premium adhesives, and in industrial flooring.

Silicones are used for protective wall coatings against humidity.

ABS and PP are used as pipes and accessories. Polybutene and cross-linked PE are utilized for hot water applications. Chlorinated PVC is also used for this purpose, however, it is sensitive to impact.

There are polymeric additives for concrete, bitumen and asphalt.

The difficulties for massive penetration of synthetic plastics into the building industry can be summarized as relatively high price, lack of adequate engineering data, lack of sufficient information regarding weatherability, and the danger of fire. Other drawbacks are creep under load, insufficient mechanical properties, and the conservative attitude of both builders and tenants. Know-how and experience have been accumulated with time, so that currently specific engineering data about synthetic plastic building materials (including long-term creep) are becoming available.

As for weatherability and external endurance, accelerated tests in the laboratory (as well as practical evidence from severe climatic regions) lead to life expectancy of about 15 to 20 years, for rigid PVC and reinforced polyester. Adequate stabilization should obviously be applied when these polymers may be sensitive to heat (above 60°C), mainly when dark colors are used. As for the danger of fire, PVC tends to extinguish itself, while other polymers require special anti-fire reagents. On the other hand, the appearance of smoke and toxic vapors during fire must be considered. It should be emphasized that the utilization of plastics necessitates engineering design that fits the specific properties of the material. It is mandatorily forbidden to use designs that fit conventional materials for plastic materials.

The optical properties of acrylics, PC, PVC, PS and PES provide the architect with many decorative options. Coating with fluorocarbon polymers improves weatherability and the use of PC is growing fast. Prefabricated structures for housing, military and development areas utilize polymers in various ways as reinforced laminates or *in situ* casting of reinforced polyester with insulation of foamed PUR. The use of PS, PUR and UF foams may save energy for heating (or cooling). Decorative sheets are made directly of synthetic polymers or wood impregnated and glued with polymeric resins. Foamed plastics as frequently used in a rigid sandwich and structural foams are currently replacing wood in furniture or other structures. Consequently the prospect of expanding the use of synthetic materials in prefabricated construction is very promising. A breakdown of the use of plastics in U.S. construction is given in Table 6-3.

Uses in Packaging   About one third of all plastics appear in various forms of packaging, either as containers or as films and bags — about 50% in containers and bottles, 38% in films and bags, and about 8% in coatings. Among the polymers that appear in packaging the following should be mentioned — polyethylene — LDPE (33%), HDPE (28%), PS (12%), PP (10%), PET (8.5%), PVC (5%). The rest consist of polymers like PVA, Nylon, cellulose, PC, SAN, PVDC, and EVOH.

In containers and bottles, the most prominent polymers are HDPE (milk bottles) and PET (bottles for gaseous soft drinks). Rigid PVC is found in bottles for mineral water and non-gaseous drinks, but its use has been drastically reduced (particularly for food) as a result of health hazards (including difficulties in recycling). Another distinguished group is comprised of bags and sacks for food and industrial goods. In this case, flexible LDPE is advanta-

TABLE 6-3
Use of Plastics in Construction
(United States, 1990)

Use	Percent
Decorative elements	1.0
Flooring	4.7
Glazing	2.4
Insulation	10.5
Lighting	1.2
Panels	7.5
Piping	39.3
Sanitary appliances	1.7
Profiles	3.5
Adhesives for wood	25.2
Vapor sealing	1.8
Wall coating	0.9

geous while the new type LLDPE (with improved barrier and mechanical properties) appears as strong competition. Other polymers like HDPE (including UHMW), PP and plasticized PVC also are used as films with thickness normally below 0.25 mm. For stronger polymers, the most common thickness is around 0.075 mm. Polymers may also appear as a coating on cardboard or aluminum packaging (mostly LDPE) or as the interior layer in wooden or metallic containers.

The requirements for films in packaging are mainly tensile and tear strength, elongation (flexibility) and impact strength. In addition, attention should be given to low permeability to water, vapors and gases, as well as chemical and thermal endurance. Polyethylene has also the advantage of easy sealing by welding. It replaced mainly paper and cellophane, but the latter are still found mostly in automatic packaging machines, providing high light transmission, gloss and mechanical strength, but unfortunately absorb and transmit water vapor and are unweldable. In general, by coating and lamination (currently mostly by coextrusion) multilayer films are formed, so that only the external faces must be protected against water diffusion and provide easy weldability.

The advantages of polyethylene stem from high flexibility and its barrier to water together with high permeability to air and gases, thus providing "breathing" for fresh food stuff and eliminating water losses. On the other hand, when sealing for gases and odors is essential, polymers like PVC, but mainly PVDC, EVOH (copolymer ethylene-vinyl alcohol) and Nylon are called for. PP (usually after stretching to include orientation) serves in films and tapes of premium mechanical strength and higher thermal resistance (compared to PE). HDPE too has penetrated the market for rigid grocery bags — a substitute for paper.

Quite strong films are made of saturated polyester (mostly after biaxial stretching) which endure higher temperatures. Multilayer films, as mentioned before, are offered for specific applications wherein the barrier properties of one polymer are exploited, together with another polymer that shows water repellence and easy weldability. These multilayer films are produced by lamination but mainly by coextrusion into 5 to 7 layers. Interesting combinations are polyester and Nylon, PE and EVOH, for extended shelf life. EVOH excels as a vapor barrier, but is very sensitive to water, so that it has to be shielded by PE. PET appears abundantly in films, exhibiting a good barrier to water and vapor, transparency and resistance to high temperatures. The barrier properties may be improved by a laminate PET/PA/PET. CPET (crystalline PET), an opaque material, serves in trays that may also be heated with food by microwave. PAN (acrylonitrile) or its copolymer with acrylate (Barex™) excel as a barrier to vapors. In vacuum packaging of meat or cheese, the combination of layers of LDPE or EVA is recommended. Excellent barrier properties are found in the novel polymers, LCP, in aromatic polymers of very low thickness; however, their high price is still a limiting factor for common use. Quite interesting are shrink films made of PE, EVA or PVC, and films that withstand boiling water.

In rigid packaging (mostly dairy products) the most common polymer is PS (mainly HIPS or expanded EPS). For single use containers and bottles the common polymers are HDPE, PP, PVC and PET. The innovation is connected to biaxial orientation thus obtaining higher strength but mainly a barrier for vapors (like PET). Transparent bottles of high toughness are provided by PC, while bottles that withstand high temperatures for sterilization are made from PBT. Another packaging item consists of various boxes for bottles, food stuffs and industrial goods, made mostly from HDPE, PP or rigid PVC. Insulating packaging is widely used, mostly with polystyrene (EPS). The main issue with food is fear of toxicity from monomer traces, mostly in the case of VCM. Another problem is concerned with ecology of single use packaging which should be directly recyclable, or at least indirectly reused for energy or chemical derivation.

Incidentally, the replacement of commodity packaging like glass, aluminum or paper originated not only because of the unique properties of plastics (including low weight) but also because of the relatively low energy requirement. For example, the energy requirement for the formation of an aluminum container is around 3 KWH (including production of the metal) and 2.40 KWH for glass, while a plastic container of similar volume requires only 0.1 KWH, which means that the saving in energy is remarkable.

Uses in Transportation   The automobile industry presents an enormous challenge for plastics, due to the huge production capacity in this field. The various options for replacement of common materials like metals or glass by premium synthetic materials would contribute a significant reduction to the weight of the vehicle (resulting in a real saving of fuel), enhancement of corrosion resistance, and variety in composition and design. While the use of plastics in a vehicle was minor in 1960 (around 10 kg per car), it has since increased enormously—about 6% of the weight of the car in 1980 (about 90 kg), 8% in 1985, and currently about 10%–15% (meaning 100–180 kg per car). In the U.S., the use of plastics in cars during 1990 amounted to over 10^6 tons, not to mention the enormous use in ships and aircraft.

In the early years, the major plastics used consisted of unsaturated polyester reinforced with fiberglass. Altogether hundreds of items in the car are made from plastics, including insulation for cables and piping. The most apparent are polyurethanes, either as flexible foams (for seats) or as a rigid foam for insulation, in many cases processed by a reactive injection process (RIM). PUR is also available as an external coating. PP (mostly reinforced) is increasingly found in combination with foamed elements for the bumper, in which case a polyblend of an olefinic polymer and elastomer is applied for enhancement of toughness. PP is also used in the production of battery containers. PVC is still used in the insulation of wire and cable or upholstery of seats. ABS is also much in use for many items, some coated by chromium or nickel. In racing cars, ABS has successfully replaced steel in the main body. Many other construction parts in the car will be gradually replaced by composites like epoxy–Kevlar, epoxy-graphite and others.

Many appliances are already produced from engineering polymers, while engineering polyblends such as Nylons, acetal, PBT, PC and Noryl are becoming a very popular. Polycarbonate is much used in lighting fixtures due to its high light transmission and excellent toughness, but PMMA is also available for this purpose. Nylon is found in fuel tanks, handles, switches and a variety of other items. The steering wheel and other equipment under the hood are a festival of plastics. Both linear and branched PE, Bakelite, and reinforced polyester are utilized. In conclusion this story is by no means exhausted. Recycling of plastics has become vital and the industry currently pays great attention to ecological problems. (See Chapter 7.)

In cooling or commercial vehicles, plastics for insulation and storage have become essential. Add to this the enormous use of synthetic elastomers for tires. A breakdown of polymers used in the transportation industry is shown in Table 6-4.

Reinforced plastics (unsaturated PES) are abundantly utilized for maritime vehicles like sailboats or racing boats, as well as commercial boats. In aviation and space, construction elements (sandwiches) are increasingly applied. They are mainly made from epoxy, reinforced by specific fibers, silicones and high-performance polymers like poly-ether-imide, PEEK and others. Sophisticated polymeric systems that exhibit extraordinary performance show promising potential in these fields. Their major contribution consists of low-weight, inertness and stability.

Uses in Medicine   The application of synthetic polymers in medicine is relatively new, causing a real revolution. It covers medical aids, accessories,

TABLE 6-4
Use of Plastics in Transportation

Polymer	%, Use
PUR	21–24
PP	18–20
PVC	11–18
Reinforced PES	9–12
ABS	6–11
PE	4–6
PA (Nylon)	3–7
PMMA (acrylic)	2–3
POM (acetal)	2–2.5
PC	1–2

implantations and artificial limbs. In addition, polymers are also utilized as carriers of self-controlled medicines (in microcapsules). When polymers enter the human body, purity and chemical stability are of major concern, most important being biological compatibility (avoiding rejection). This calls for a test of medical grade. In general, the following applications are of importance — conduits for blood delivery, heart valves and pacemakers, pumps, injectors, elements of artificial kidneys, pancreas and heart, dialysis and infusion systems, limbs, synthetic casts, catheters, threads for surgery, glues for bones, artificial joints, contact lenses, infusion bags, and artificial teeth.

Other uses are found in plastic surgery, artificial skin and blood substitutes. An unique field is found in membranes for dialysis. PE, mostly linear (including UHMW), serves in many implants or in artificial joints. PVC is the most useful polymer in medicine, in plasticized form for flexible tubing, dialysis and infusion systems. PP is offered for disposable injections and, together with ethylene (copolymer), in blood bags. PS serves as a substitute for glass in tubes, bottles and petri dishes, and as the copolymers SAN or ABS in a wide host of uses. Polyester (PET) is used in sewing threads or nets in prostheses. PC is used (as a replacement for cellophane) in membranes for dialysis as well as in blood pumps and other systems.

Another engineering polymer, polysulfone, presents excellent resistance to sterilization while it is also well accepted by the human body and thus is adequate for implanted systems.

PMMA is mostly utilized for vision — both hard and soft lenses (usually as copolymers), implanted lenses, and artificial eyes.

Silicone is widely used in artificial limbs and in fixing defects of the body, membranes, implanted capsules for slow-release medications, artificial skin and in insulation of implanted electrical devices.

Polyurethane is used in similar ways, including an artificial heart made of polyether-urethane.

Special items are made of Teflon, Nylon and acetal.

Widely specific packages are in use, like shrink and sensitive systems. The polymer industry faces an increasing challenge and use in the field of medicine.

6.3 POLYBLENDS AND ALLOYS

Polyblends and alloys (compatible systems) introduced a new dimension to the world of polymers, leading to a significant saving by avoiding the manufacture of new polymers or copolymers. These are industrial mixtures that combine certain properties and thus often reduce the cost of the product. When compatibility is reached, division into separate phases is eliminated (which may reduce mechanical properties). Therefore the use of compatibilizers as coupling agents has been increased. In the ideal case of compatibility between polyblends, there appears a single transition temperature T_g, as opposed to the individual transition contributions which must be dealt with in a regular polyblend.

Polyblends are mostly used in the domain of engineering polymers. The first and most prominent one is Noryl (a blend of PPO with styrene), where styrene contributes workability and economics. Polycarbonate appears in many combinations (PC with ABS, PC with PBT). In 1987 the use of polyblends already reached 250,000 tons in the United States, comprising one third of the consumption of all engineering polymers, at a rate of growth 2 to 3 times that of regular polymers.

The major use of polyblends is in transportation (50 to 60%); the rest is in domestic and office appliances, and electronics. The following provides a partial list of useful polyblends:

PPO/HIPS (Noryl)^{TM} offers a combination of toughness, endurance at high temperatures and improved workability.

PC/ABS (Cycoloy or Bayblend) has the enhancement of low temperature impact strength, improvement of workability and favorable economics.

PC/PBT or PC/PET alloys (Xenoy) combines high impact strength and resistance to chemicals.

PC/SMA (polycarbonate with styrene-maleic anhydride alloy) exhibits toughness and endurance at high temperatures.

Other useful combinations are ABS/PVC, PP/EPDM or polyblends which improve properties of Nylon, polysulfone and acetal.

6.4 PRODUCT DESIGN

The availability of a wide choice of polymers and plastics enables optimal design of products. Due to the large number of families of polymers (each family widely subdivided, with or without reinforcement or modifications, not to mention polyblends), it is not surprising that choice of the proper material is rather complicated. While the computer may be used to provide data about material properties (and flaws) and to direct the correct decision, there is a great deal to know about the available materials and their performance.

The major issue is economics, while the quality of the performance is always decisive. The method of processing is an important factor in the product design and final shaping. The option of recyclability should be taken into consideration. The designer must be warned against copying designs that are common in conventional materials like metal, glass, wood and others.

At an initial stage, the purpose and required physical properties of the product must be carefully defined as mechanical, electrical, optical, chemical, aesthetic, as well as resistant to environment or fire. When performance is involved, long-term behavior must be examined. Accordingly, the choice can be applied by eliminating several materials that fit all the required properties in order to select the economical one which combines cost of materials and processing. Processing itself may offer several competing options. Regarding the cost of raw materials, the price per volume is dominant, though the actual price is ordinarily given per weight.

When high performance is required (like in metal substitutes) the engineering polymers are the appropriate choice, wherein, polyblends often provide a compromise between material cost and the basic properties. In the field of aviation, space and military, specific polymers are used providing excellent endurance at a wide range of temperatures. In this case the quality of performance dominates economic considerations (in spite of a permanent concern about economy). As mentioned before, quite complicated and exact items may be obtained by injection molding. Major sophistication rests in mold structure, so that all elements (metallic insets, several layers) can be simultaneously introduced. It is essential to reach the final shape without the need for additional processing (like painting or milling). Currently, use of the computer is dominant. It is the CAD (computer aided design) which selects the various elements of the process and the controlling parameters of production quality and rate. Sharp corners should be avoided, as well as, narrow ends or drastic changes in thickness. In critical cases the stage of annealing is important. Improper processing like undesired orientation, residual stresses, unplanned contraction and flow marks can produce flaws which must be avoided.

6.5   FUTURE OF THE PLASTICS INDUSTRY

According to early expectations, the use (in volume) of plastics has already surpassed that of metals in the 1970s, while annual growth rate reached 3%–4%. During 54 years (from 1940 to 1994) the production capacity of plastics in the U.S. increased from 45,000 tons to 36 million tons per year. About 10^6 people are employed in this industry in the U.S., while about 60% of all engineers and chemists are involved in polymers. Global consumption has reached about 110 million tons (money value of $200 billion), while the expectation for the year 2000 is around 170 million tons.

Major growth is anticipated in the field of engineering and high-performance polymers, composites and polyblends. Composites consist of engineering polymers reinforced with glass fibers, graphite or ceramics that may replace metals in various outlets such as automotives and transportation, machinery and electronics. In 1994, the global consumption of engineering polymers reached a bulk of 7.5 million tons (including ABS and SAN). It is expected that about 20% of the metals in use will be replaced by the year 2000 by engineering polymers. Among novel polymers LCP and PEEK are of great interest, as well as electrically conducting polymers and others with unique performance. An internal combustion engine has already been made of polyamide-imide, and about two thirds of the weight of all aviation vehicles will be made from composite polymers by 2000, while the use of plastics in space, automotives and construction will expand.

The processing of plastics has moved to well developed methods like CAD and CAM, which improve control and automation in production. Special 3-D (three-dimensional) CAD is available in many aspects for production of molds, and design of products up to final production. This design is based on software that consists of data on physical properties and condition of processing for a wide number of polymer types. Table 6-5 summarizes properties and utilities of selected polymers. Tables 6-6 through 6-10 summarize the chemical properties, thermal, electrical and general properties, relative prices and consumption of polymers in the United States and globally.

TABLE 6-5
Properties and Utilities of Selected Polymers

Polymer	Typical Properties	Major Utilities
Polyethylene, LDPE	lightweight, flexible, inert towards water and chemicals	packaging, piping, films for agriculture, electrical insulation
Polyethylene, HDPE	lightweight, rigid, inert towards water and chemicals	rigid packaging, piping
Polypropylene, PP	lightweight, rigid, excellent chemical resistance	packaging, tools, textiles, rugs, furniture
Polybutylene, PB	lightweight, resistant to high temperatures	hot water conduits
Polyvinylchloride, PVC	stable, useful, versatile in compounding, rigid or flexible	construction, electrical insulation, furniture, upholstery, packaging, piping, flooring
Chlorinated polyether (Penton)	chemical and thermal endurance	piping and accessories for chemical industry
Polystyrene, PS	rigid, shining, easy for tinting	electrical appliances, refrigerators, toys, housewares, thermal insulation (foam)
ABS	rigid, tough, easy workability	accessories, machinery, telephones, automotives
Acrylics, PMMA	rigid, transparent, stable towards environment	lighting, opticals, glazing
Nylon, PA	mechanical strength, engineering properties	fibers, machinery, brushes, premium food packaging
Cellulose	rigid, tough, transparent	packaging, toys, commodities
Acetal, POM	rigid, stable in dimensions, high performance	machinery, sanitary accessories
Polycarbonate	high impact strength, mechanical and thermal stability, transparent	optics, electronics, machinery, lighting, glazing

TABLE 6-5
Properties and Utilities of Selected Polymers
Continued

Polymer	Typical Properties	Major Utilities
Polyester (Saturated)	rigid, thermal endurance, stability, barrier to vapors, chemical resistance	engineering and electrical appliances, bottles
Fluorocarbons	chemical and thermal endurance, stability, low friction, inert, expensive	anticorrosive sealing and coating, special piping
Noryl, PPO	rigid, tough, stable, thermal and chemical resistance	tools and machinery, automotive
Polysulfone	premium mechanical and thermal properties	engineering use at high temperature
Polyphenylene-sulfide, PPS	thermal and environmental stability, fire extinguishing	machinery and electronics, high temperature coatings
Polyimide, PI	thermal and mechanical stability, very expensive	fibers and films for high performance
Ionomer	flexible, transparent, easy workability	special packaging, containers
Phenolics, PF	thermal and chemical endurance, cheap	electrical accessories, coatings, adhesives
Urea, UF	strong and bright	coatings, adhesives, foams
Melamine, MF	strong, hard, bright	coatings, dishes
Epoxy	bonding, chemical resistance	coatings, adhesives, construction
Polyester (Unsaturated)	good strength when reinforced	reinforced plastics in construction, coatings
Polyurethane, PUR	chemical and thermal resistance	foamed plastics, coating, elastomer
Silicones	high temperature endurance, water repellant	HT coatings, elastomers

TABLE 6-6
Mechanical Properties of Useful Polymers

A. THERMOPLASTICS	Specific Mass	Tensile Strength MPa	Compression Strength MPa	Flexure Strength MPa	Tensile Module MPa	Elongation %	Impact Strength kgm/m	Hardness Rokwell
Polyethylene, LD	0.92	12	3		120	400	80	R 11
Polyethylene , HD	0.95	20	16	10	1000	300	40	R 30
Polypropylene, PP	0.90	33	63	46	1100	500	30	R 100
Polystyrene, PS	0.95	40	73	85	3300	12	1	R 80
PVC (rigid)	1.40	33	85	80	5300	10	2	M 70
Acryl (PMMA)	1.20	100	100	30	2700	5	2.5	M 100
Nylon 6-6, PA	1.14	85	73	70	2700	75	5	R 115
Cellulose acetate	1.2	30	140	70	1400	20	20	R 100
Acetal, POM	1.43	66	60	100	2700	25	7	R 120

TABLE 6-6

Mechanical Properties of Useful Polymers

Continued

A. THERMOPLASTICS (continued)	Specific Mass	Tensile Strength MPa	Compression Strength MPa	Flexure Strength MPa	Tensile Module MPa	Elongation %	Impact Strength kgm/m	Hardness Rokwell
ABS	1.02	43	73	70	1800	70	35	R 96
Polycarbonate, PC	1.2	60	73	80	2100	80	70	R 70
Teflon	2.1	12	11		400	150	20	R 20
PET (oriented)	1.4	160			4100	180	10	R 105
PBT	1.31	55	77	100	2500	200	5	M 72
B. THERMOSETS (reinforced)								
PF	1.85	50	140	200	20,000	0.2	75	M 100
UF	1.5	60	200	100	7000	0.6	2	M 115
MF	1.5	250	470	440	10^4	0.7	60	M 120
Epoxy	1.85	320	500	470	10^4		65	M 100
UPES	1.75	100	270	160	8000		50	M 95
Silicon	1.68	150	230	430	10^4		60	M 100

TABLE 6-7
Thermal, Electrical and General Properties of Useful Polymers

A. THERMOPLASTICS	HDT °C	Max. T°C	α $1/°C \times 10^5$	K $kCa/h.m.°C$	Specific Resistance $ohm\ cm$	Dielectric Strength $kV\ cm$	Dielectric Constant (60 HZ)	Dielectric Loss Factor (60 HZ)	H_2O Absorption 24 h %
LDPE	38	93	18	0.34	10^{17}	180	2.3	0.0005	0
HDPE	52	107	12	0.34	10^{17}	160	2.4	0.0004	0
PP	105	127	14	0.42	6×10^{17}	230	2.25	0.0005	0
PS	72	72	8	0.10	10^{19}	200	2.5	0.0005	0.5
PVC (rigid)	72	65	10	0.18	10^{16}	480	3.5	0.001	0.5
PMMA, acrylics	70	88	8	0.18	10^{15}	200	4.0	0.05	0.3
PA, nylon	80	140	12	0.90	10^{14}	140	4.3	0.015	1.0
CA	65	55	12	0.15	10^{11}	80	5.5	0.01	5.0
POM, acetal	100	120	8	2.4	10^{14}	190	3.8	0.004	0.2

TABLE 6-7
Thermal, Electrical and General Properties of Useful Polymers
Continued

A. THERMOPLASTICS (continued)	HDT °C	Max. T°C	α $1/°C \times 10^5$	K kCal/h.m.°C	Specific Resistance ohm cm	Dielectric Strength kV cm	Dielectric Constant (60 HZ)	Dielectric Loss Factor (60 HZ)	H_2O Absorption 24 h %
ABS	90	120	10	2.2	10^{17}	150	3.25	0.016	0.2
PC	135	150	7	0.70	2×10^{16}	180	3.0	0.0007	0.3
PTFE	60	260	10	0.20	10^{16}	160	7.0	0.0003	0
PET	63		9	0.90		220	3.2	0.03	0.1
PBT	54	140	7	0.20		160	3.2	0.02	0.1
B. THERMOSETS									
PF	290	260	1.8	0.3	10^{12}	200	7.0	0.01	0.5
UF	140	77	2.5	0.25	10^{12}	140	8.0	0.01	0.6
MF	200	150	1.0	0.45	10^{13}	240	6.8	0.09	0.5
Epoxy	230	150	1.8	0.75	10^{16}	280	3.8	0.03	0.15
UPES	230	120	1.8	0.75	10^{14}	260	4.3		0.45
Silicon	450	350	1.0	0.25	10^{15}	160	3.9		0.3

TABLE 6-8
Consumption and Relative Prices of Commercial Polymers,
United States, 1994

Polymer	Annual Consumption 1,000 tons	Consumption Percent of Total	Relative Price
LDPE	6,400	17.8	1.0
HDPE	5,285	14.8	1.0
PP	4,437	12.4	1.0
PVC	5,056	14.1	0.7
PS	2,669	7.5	1.3
ABS	680	1.9	1.8
SAN	59		1.8
POM	97		2.5
PA	418	1.2	2.9
PC	317		3.5
PPO	109		3.0
PET	1,568	4.3	2.0
UPES	1,335	3.7	1.4
PF	1,465	4.1	1.0
MF, UF	994		1.5
PUR	1,708	4.8	1.8
EP	274		2.6
Total:	35,770		

TABLE 6-9
Physical Properties and Consumption of High Performance Polymers (United States, 1993)

Polymer	Specific Mass	Tensile Strength MPa	Rigidity Modulus MPa	Tensile Elongation %	Impact Strength Kg.m/m	HDT °C	Global Consumption 1.000 t	Relative Price*
Polysulfone, PSU	1.24	68	2,400	50	9	174	10.2	8.5
Polyethersulfone, PES	1.37	80	2,400	60	8	200	2.4	10–25
Polyphenylene sulfide, PPS	1.67	120	10,000	1.5	6	260	14.2	3–6
Polyarylate, PAR	1.20	63	2,000	50	2.3	174	2.4	5–6
Polyether-imide, PEI	1.27	100	2,800	7	5	200	1.4	10
Polyamide-imide, PAI	1.40	180	4,300	12	1.6	260	0.3	40
Polyether-ether-ketone, PEEK	1.32	90	3,500	50	9	160	0.3	50
LCP	1.70	100	12,500	2	9	355	2.7	20

*The basis 1 is for commodities (LDPE at 44–50 cents/lb)

TABLE 6-10
Global Consumption of Polymers, 1994

Location	Consumption 1,000 tons
United States	37,315
Western Europe	31,065
Eastern Europe	6,575
Latin America	3,830
Asia	28,620
Mideast	2,100
Total	110,000

PROBLEMS

1. Describe the milestones in the development of the polyolefins.

2. Describe the differences in molecular structure, methods of polymerization and general properties between low-density (LD) and high-density (HD) polyethylene.

3. What is the significant contribution of the stereospecific polymerization process to the development of polypropylene?
   Compare PP to PE.

4. What is the uniqueness of LLDPE?; UHMWPE?

5. Describe structure-property interrelation in the case of polystyrene.
   How about morphology?
   How do you improve its toughness?
   How do you expand it?

6. Describe and compare modification of vinyl-chloride by copolymerization or by plasticization (additives).

7. What properties are required from engineering polymers?
   List and describe some of them.

8. Describe the Nylon family, and discuss advantages and disadvantages. Compare aromatic to aliphatic Nylons.

9. Describe polyblends and alloys.
   What do we gain?

10. Describe synthetic elastomers and compare to natural rubber.
   Define the term elastomer.
   What is the importance of thermoplastic elastomers?

11. Describe synthetic fibers and compare to natural fibers.

**There is a wide world of plastics—
Here are a few examples
of various products**

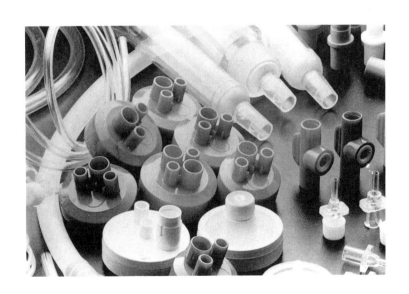

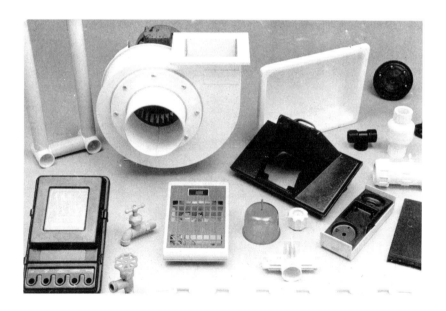

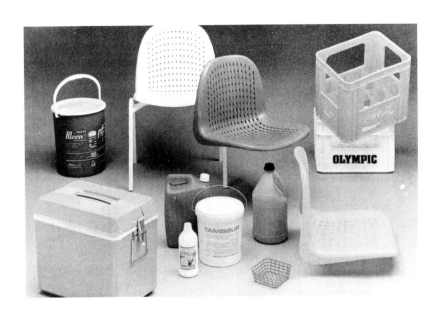

CHAPTER 7

Plastics and Ecology

PLASTICS AND ECOLOGY—are these in conflict with each other? Apparently, plastics are quite friendly to the environment—they are not hazardous to nature, they do not contribute any essential contamination to the air, water, or ground. We may regard these materials as nonpollutants. Yet there appears to be an increasing bias against plastics in public opinion, due to the fact that more and more plastic packages are observed in litter or in domestic wastes. The problem of litter, can be solved both by regulation and education. As to the growing appearance of plastics in domestic (as well as industrial and agricultural) waste streams, in contrast to other organic components (including paper), plastics are stable towards biological attack, and have a very long lifetime. Now, plastics do not represent the largest component in domestic wastes, yet as containers and packages they are bulky and therefore consume relative large volumes. In conclusion, the conflict between plastics and ecology may be confined to the accumulation of plastics (mainly from packages) in landfills that are becoming crowded and scarce. The treatment of domestic wastes in an economical way has become a major issue, while ecologists fight for a clean and pleasant environment. The various solutions and a survey of the state of the art will be discussed.

7.1 REUSE, REDUCE, REPLACE

What are the various options for reducing the amount of plastics in municipal solid waste? The first solution increases the life cycle of plastic by multiple use (refill) rather than disposal of packages. This is practiced with glass bottles via collection and a cleaning process. The latter, however, uses detergents, which eventually are discarded to the sewage streams and add contamination to the environment. (Actually, even glass bottles are mainly disposed of nowadays.) In the case of plastic containers, this is not an economical nor an ecological solution, except for very large containers.

Source reduction in packaging is possible to some extent, as overpackaging is not always necessary. The size of some packages can be reduced without any real losses. (Quite the contrary, it is even in favor, as it also reduces the total expenses.) In other cases, replacement of one plastic material by another type or grade that shows improved mechanical properties can lead to thinner films or containers, thus reducing the amount of disposable material. Modification of design can also contribute to size reduction. This solution however leads only to marginal results, as packaging is an essential marketing item in shipping, handling and display.

The idea of replacing plastics by other conventional materials has been suggested in many countries, mainly by ecologists. In some cases, legislation has been passed to ban plastics from the packaging market. This campaign has failed, as it was based on a misconception and even ignorance. It is easy to refer to synthetic materials as unnatural or unfriendly. But is it really true? Can the world survive and progress without man-made materials? It is not only the case of packaging, but textiles are also a great consumer of synthetic materials. Now, in packaging, plastics have replaced many conventional materials (paper, glass, metals) due to improved properties—low weight, resistance to moisture, inertness, safety, easy processability and viability. Returning to older means of packaging will not solve the problem of the reduction of solid wastes. On the contrary, larger packages will appear. If more paper were used instead of plastics, more trees must be cut. Is this good for the environment? Comparing the energy required for the production of plastic packaging to that of paper, glass or metals, the advantage of plastics is apparent. It is not only an economic issue, as more fuel for energy creates more emissions to the environment.

Therefore, taking into account all aspects, overall environmental damage due to plastics is less detrimental than for conventional replacement materials. An illustrative example will highlight this issue, that is, replacing expanded polystyrene (EPS) cups by uncoated paper ones. The weight of the EPS is just 1/6 that of the paper, while the paper consumes 12 times as much steam, 36 times the electricity and 2 times the amount of cooling water. In addition, the volume of waste water for the paper process is 580 times that of the EPS process, and residual waste water contaminants are 10 to 100 times larger for the paper production. While emissions to the air are 23 kg/ton for bleached pulp versus 53 kg/ton for EPS, one should take into account the fact that the weight of the same number of paper cups is 6 times that of EPS. Last but not least the price per paper cup is 2.5 times greater than EPS.

Comparing another common polymer, polyethylene (PE) used for grocery bags as compared to kraft paper reveals the followings—PE sacks use 20% to 40% less energy than the paper, create 74% to 80% less solid waste by volume and have 63% to 73% less atmospheric emissions. Both PE and PS resin production processes emit fewer particulates and less NO_x, SO_2 and CO than the paper processes. They also give less total waterborne waste of dissolved solids, suspended solids, acids and oxygen-depleting effluents than paper. Needless to say, in most cases, plastics replaced paper bags and cups due to improved performance and convenience, so that returning to the older packaging materials is not probable.

Even in the case of disposable diapers, returning to older cloth diapers is not likely because they create 10 times more water pollutants and use 3 times more nonrenewable energy. And what about convenience and health? Since it is still better for the environment to use disposable plastic diapers rather than reusable ones, there is hope that in the future disposable diapers will be also biodegradable. Comparing the energy requirements for plastic containers to those of aluminum or glass (at the same volume) is even more striking—0.1 KWH for PE as compared to 3 KWH for aluminum and 2.40 KWH for glass. Milk is also sold in PE bags (pouch) which if compared to glass bottles, the saving of weight, energy and heat content is remarkable. (Details are given in Table 7-1.) In conclustion, the replacement of plastics in packaging is impossible.

7.2   RECOVERY AND RECYCLING

Let's consider the various modes of recycling:

Primary recycling
Secondary recycling
Tertiary recycling via chemical treatment
Recovery of energy

Quite another approach deals with the conversion of plastics wastes to self-degradable species. This idea will be discussed later.

Primary Recycling

Primary recycling means reuse of off-grade and scrap materials directly by the converter plant. This refers strictly to thermoplastics, which already

TABLE 7-1
Comparison of Energy Cost of Manufacture for Disposable and Returnable Milk Containers

Container	Weight	Energy Used	Heat Content
Two-quart glass milk bottle	37.1 ounces	8.36 KWH	0
Two-quart plastic (PE) pouch	0.97 ounces	0.84 KWH	317 kcal

amount to about 80% of all polymers in use. During processing there is always some percentage (as much as 20% to 30%) of scrap (such as runners and sprues in injection molding, trimmings in blow molding and extrusion) as well as off shapes or products that did not pass quality control. This material is only slightly (if at all) contaminated, but has undergone some thermal history. In most cases it can be reground and added as a blend to virgin raw material (usually 10% to 25%). This is the common practice in industry, and presents an approach of good housekeeping. In some cases, however, because ultra-strict specifications do not allow regrind to be used, it will be sold to other converters. Therefore, primary recycling of homogeneous and well-defined plastics is the easiest solution, if it is possible and viable.

Secondary Recycling

Secondary recycling means that used products are collected, cleaned, often separated, and finally reused (after grinding) either as separate species or in a mixture (blend). As long as a single polymer is dominant, the option of recovery is promising, albeit the uncertainty of economic feasibility. However, industrial secondary recycling will succeed only if there exists a strong motivation, either through legislation or economically. Recycled plastics are always inferior to virgin ones, and can only compete in periods of scarcity of raw materials or temporary high prices of the latter. In most cases the recycling industry must be subsidized by local authorities who endeavor to decrease the bulk of wastes in landfills. An integrative process is called for, as plastics are never the dominant component in domestic wastes. In any case, the separation of plastics from other waste components, in addition to the separation of individual polymers within a blend of several plastics, is neither a simple nor a cheap process but technology has developed so that in principle such separation processes are becoming possible. To understand the role of plastics in municipal solid-waste streams, let us consider the typical statistical data of Table 7-2.

The weight concentration of plastics differs from country to country, and may reach as much as 12%. The volumetric concentration is much higher for plastics that are light in weight and often discarded as bottles and containers of very low bulk-density. So the volumetric concentration may reach 16% to 24% and more. The potential of recycling viable materials in general and plastics in particular will be highlighted if we look at the typical size of the municipal waste (MSW) in developed countries which is around 1.5–1.8 kg per capita per day! Thus the total amount of MSW in the United States had reached 180 million tons in 1990, of which only 13% was recovered, 14% was used for energy, while the majority (73%) was still dumped into landfills. This means that enormous amounts of solid wastes are accumulated which may lead to environmental hazards unless successfully treated. Yet, the question should be asked—is it at all possible and practical to treat these vast wastes in the economy of the free world while, at the same time, vacant

TABLE 7-2
**Breakdown of Typical Domestic Wastes
in the United States around 1990**

Material	Percentage of Total Waste
Paper	37.1
Yard waste	17.9
Metals	9.6
Glass	8.6
Food waste	8.1
Plastics	7.2
Others	8.6

land for collection of wastes is diminishing and the costs of domestic waste treatment reach the gigantic amount of 100 to 150 dollars per ton.

Recyclers confront many obstacles while trying to sustain sound industrial concepts. Their first concern must be the purity and consistency of available sources. It is always advantageous to avoid separation processes, so that homogeneous sources are naturally favored. Such sources for recycling plastic wastes appear in agriculture (coverage of greenhouses and tunnels, irrigation pipelines and accessories) and in industrial or post-consumer wastes (mainly used packages). The major contributor to wastes is the packaging industry which happens also to be the major outlet for plastics (about 30%). A typical breakdown of various polymers that appear in packaging is shown in Table 7-3. The role of PET is increasing while that of PVC is diminishing in many countries.

There are other short-term outlets for plastics use in medicine, transportation, toys, etc. A breakdown of plastics use in various utilities in the United

TABLE 7-3
Breakdown of Polymers Used in Packaging

Polymer		Percentage of Total Weight
LDPE	(low density polyethylene)	33
HDPE	(high density polyethylene)	31
PP	(polypropylene)	10
PET	(polyethylene terephthalate)	7
PS	(polystyrene)	5
PVC	(polyvinylchloride)	5

States in 1990 is shown in Table 7-4. Other countries use more plastics in agriculture.

It is clear that many polymers found in packaging or other short-term uses, are commodities, relatively cheap even for virgin raw material. Only PET belongs to the upper-class of commodities which leads to an increasing interest in reutilization. When speaking about secondary recycling of plastics, it means that the collected material must be identified as to its polymeric generic family (including additives) as long as a single component is recoverable. After cleaning and grinding, it must be roughly characterized and directed into final products that tolerate less restrictive demands. Therefore, imaginative design and marketing enterprise is the key to success. Actually, each polymeric family has its limitation and merits, while a mixture of two or more polymers (not mentioning traces of other materials) may often lead to weak and low-grade products.

A technical term in polymer science is the ill-defined idea of compatibility. Basically incompatibility is the rule, meaning that most polymeric blends tend to separate into their individual phases, resulting in the weakening of their cohesive bonds. The exception lies in so-called "alloys," or compatible blends, as previously described in the discussion of polyblends. Unfortunately, among those polymers that appear in domestic waste, compatibility is seldom met. So, how can domestic waste be treated in order to follow ecological regulations? The first step is source-separation, which means that the domestic waste has already been separated by the inhabitants into several distinct groups—metal, glass, paper, plastics and organics. Such source-separation is already required in some countries (Germany, Japan and others) or accomplished in many cities around the world on a regulatory or a voluntary basis. The next steps consist of an industrial collection and separation center that by using modern equipment of sorting and cleaning leads to final

TABLE 7-4
Utilization of Plastics (United States, 1990)

	Percentage of Total
Packaging	30
Construction	24
Household	5.5
Transportation	5.4
Electronics	4.7
Furniture	2.4
Toys	1.5
Medicine	2

individual products. This is a complex operation, as each species must be further subdivided into various metals, glass of different colors, and finally plastics of different families and grades (obviously a noncompatible blend). Without source-separation the task of waste recycling would be even more complicated and costly.

Another approach is source-separation of homogeneous resins that are easily identified as paper, aluminum cans, clear glass and (last but not least) individual plastics. Plastics cannot always be identified by the general public as to their generic family because most plastics look similar to each other, in contrast to most metals. However, if a single polymer dominates a special field (like PET bottles for soft beverages) these can easily be collected unmixed with other types of polymers. (The supermarkets mainly promote this type of collection by repaying deposit money.) Large amounts of identifiable plastics are utilized in agriculture (mainly polyethylene as LDPE), which makes secondary recycling a simple procedure. The idea of secondary recycling is that wherever possible individual unblended plastics are preferred. Recently codes for identification have been applied for several polymeric resins. The recovered and cleaned polymer serves as an optional raw material, either as a blend with fresh (virgin) material, or directly converted to final products that tolerate lower grade quality.

If separation of a blend of plastics is not feasible, there are ways of fabricating via fusing and compressing into rough molds. These articles are usually termed as polywoods, resembling lumber in their texture, properties and uses and they are usually cheap, dark-colored, bulky and not very attractive. The real concept of recyclable materials (or recyclability) led to new designs, wherein a uni-resin product is more advantageous than items made by a combination of several resins. A good example is the popular PET bottle for soft drinks that succeeded in replacing the old glass bottles, due to a host of advantages—light weight, safe from breakage and less energy consuming. However, an early design of cups made of HDPE for mechanical stability caused some difficulty in the recycling process, calling for initial separation of the cups from the bottles. Later improved design ended with a stable bottle made of PET only and thus eliminated the tedious and costly separation stage. The automotive industry, being aware of environmental requirements, also considers a switch from polyblends to uni-resin materials. The most recycled and reprocessed plastics at the moment are PET, HDPE and LDPE. (Details about the recycling procedures and utilization of various polymers will be described separately.)

7.3 TERTIARY RECYCLING

By chemical treatment of scrap plastics, some useful products may be gained, such as monomers, oligomers, chemicals, or fuels. The method of tertiary recycling may sometimes present the only viable solution. Depoly-merization (the decomposition of a polymer into its basic reagents, mono-mers) has been explored for many years as part of degradation studies.

Controlled depolymerization (as in pyrolysis) renders selectively pure mo-nomers. In the case of PMMA (acrylics) it is successfully commercialized, and the yield is quite high. The pure monomer may replace virgin raw mate-rial. PET is also prone to depolymerization (via hydrolysis) as are other poly-condensation resins, such as polyamides (Nylon) and even polyurethanes. Whenever technology exists, the advantage of tertiary over secondary recyc-ling stems from achieving pure monomers that can eventually be converted to premium grade resins. However many polymers degrade during pyrolysis (PS, PE, etc.) and for those that undergo controlled depolymerization, the existence of impurities may hamper the process. PET has been the target of most depolymerization enterprises, due to the advantage of obtaining pure monomers for repolymerization over the secondary recycling of contami-nated or colored products. Tertiary recycling enables the use of recovered PET in food packaging, which does not accept reground secondary recycles.

Some new technology is being developed by which mixed plastics may be depolymerized at high temperatures into monomers that can be easily sepa-rated and repolymerized. Pyrolysis is also effected on commingled plastics, as well as on thermosets, wherein the product consists of a fuel, that will be used for energy. In summary, the concept of tertiary recycling may be practi-cal for some polymers while also effective for better recycling of mixed plas-tics or thermosets.

7.4   ENERGY RECOVERY—INCINERATION

Whenever separation and recycling are not feasible, or there does not exist a promising market for recycled products, a substitute solution is based on recovery of the caloric value of polymers. As hydrocarbons the energy content of commodity polymers is more than twice that of coal or paper and four times that of general waste (MSW). This is exhibited in Table 7-5.

In spite of some economical advantages, the idea of incineration is not accepted everywhere, the major problem being environmental hazards gener-ated during combustion—the toxic gases and contamination of heavy metals (acid rain), dioxins and chlorine chemicals (mainly from PVC). Wherever incineration is used, plastics are not usually separated from the total waste, which reduces the energy recovery, and also creates more ash. Japan leads in the use of incineration of MSW at about 50% (including 67% of plastics wastes), compared to 30% in Europe and about 15% in the United States.

The success of incineration depends on modern design in particular for high efficiency, and good control of pollution emission (which can cost about 20% to 30% of the whole process). However, whenever landfill is scarce (as

TABLE 7-5
Caloric Value of Plastics and Other Materials

Material	BTU per Pound
PE	19,900
PP	19,800
PS	17,800
Rubber	10,900
Newspaper	8,000
Leather	7,200
Corrugated paper boxes	7,000
Textiles	6,900
Wood	6,700
MSW	4,500–4,800
Yard wastes	3,000
Food wastes	2,600
Fuel oil	20,900
Coal	9,600

in Japan and many populated cities in the western world), incineration is becoming more attractive. In any case, scrubbing and treatment of toxic emissions are required, dealing with nitrogen oxides, chlorinated dioxins and furans, sulfur dioxide, hydrogen chloride and heavy metals. It is still believed that PVC is the major contributor to toxic emissions, which is one of the reasons that the role of PVC in disposable packaging has been diminished. Solid waste can also be added to coal in other incinerators, as long as precaution for eliminating pollutants is undertaken. By source separation of plastics, it is obvious that their high caloric value may be exploited, if burned separately.

7.5   BIO- AND PHOTODEGRADATION

The idea of self-degradation of wastes has always been considered a logical solution from an ecological point of view. It is well known that, (compared to organic wastes and paper) plastics are considered nondegradable by microorganisms. This stability is looked upon as a great disadvantage if elimination of solid wastes is desired. Therefore, the topic of rendering plastics into biodegradable or at least photodegradable materials has been challenging, but one should remember that this solves only very restricted cases such as mulching in agriculture, litter or plastic bags. The danger of self-degradable plastics is two-fold—(1) premature collapse of containers during useful lifetime or (2) diminishing properties of recycled plastics (which is exactly a contrary approach to the option of recycling).

How can a stable polymer become biodegradable? There are two examples. First, starch-loaded polyethylene has starch linkages along the polymer chain, that may be attacked by enzymes and break down to short chains that are susceptible to further bacterial degradation; in some cases polyblends with starch were offered. Second, copolymerization can take place with carbonyl groups that absorb UV radiation when exposed to the atmosphere and break down mechanically via a photo-oxidation process. It is believed that after chain scission, the shorter chains may eventually break down by microorganism attack. Similar results may be achieved by other modifications or adding controlled dosages of UV sensitizers. In other cases, water-soluble polymers are offered.

It is however claimed that, except for the commercial success in agriculture (only for the mulching of soil), the application of reducing stability in packaging has been disappointing. In most cases, no full biodegradation was achieved, other than a physical one. In addition, photodegradation calls for successful exposure in thin layers to UV radiation, which is not always possible. The thickness of the packaging plays a definite role, so that bottles and containers will be degraded very slowly. Amazingly enough, it has been found that even food and paper do not degrade well in lower, relatively dry layers of landfills, so that there is no advantage in adding degradable plastics to landfills (unless composted together with the organic material). It is clear that processes for preparing degradable polymers also add to the final cost. The mechanical properties may also be affected.

In conclusion, a solution to waste treatment by self-degradation of plastics has not been favored at least for the packaging industry. However, it has been successfully implemented for selected controlled-life plastics in agriculture (mulching). It is also a possible solution for marine debris, plastics in release control of fertilizers and pesticizers, biomedical applications, and diapers. Converting to polymers which are soluble in selected solvents may also permit some special treatment.

7.6   TECHNOLOGY OF RECYCLING

As stated before, the secondary recycling of scrap plastics from various sources appears to be the best ecological solution, as long as it is carried out according to economic criteria. Practical success depends heavily on the cost and availability of competitive options, on one hand, and on the prices of virgin raw material, on the other hand. Technology for separation and reprocessing of separated plastics have reached maturity, and recycling is being practiced in many countries already. Yet there are many obstacles, both objective and subjective. The question of public opinion has been discussed, so that we know most plastics processors already take responsibility for the time cycle of their products, including recycling. Historically, the rise of plastics recycling started in the 1970s when suddenly the prices of virgin polymers jumped, following the famous oil crisis. The boom lasted for several years only, until prices reached some equilibrium, and recycled resins could not find a proper market. However, the so-called "green revolution," that started in the early 1980s, revived and strengthened the drive towards recycling as a way of solid waste reduction. From that period, legislation has become the trigger for source-separation and recovery of major waste components, and research in these fields has flourished. The economics of alternative methods of waste disposal has been favored because the cost of handling MSW by landfilling has become very high and prohibitive. However, the route from research to industrially viable processes is rather shaky. It is obvious that an integrated process is required; namely, that in parallel, most components of MSW (glass, aluminum, paper, food residues, plastics) have to be handled and treated. The case of commercial or agricultural wastes is much simpler, since separation can be avoided and the degree of contamination is rather small.

The sequence of operations in MSW recovery is as follows: collection (source-separation is always advantageous), compacting, grinding and shredding, separating (air classifiers, water flotation, magnetic and electrostatic separators), washing and drying. The light-weight stream (after separation of paper by water slurry) consists mostly of plastics in a mixture. The most common resins appearing in MSW (mostly from packaging) are PE, PS, PP, PVC and in recent years PET.

The separation of the plastics stream is based on float–sink methods (differing densities) and, if needed, some more sophisticated ways (electrostatic, solvents, etc.) may be implemented. In many plants, hand pick-up sorting is still practiced, being (if well trained) the cheapest way. Standard coding is naturally very helpful, but in this case sorting starts prior to compacting and size-reduction. Even with the best technology of treatment and separation, failure to reach pure uncontaminated raw material, as required in industry for premium products means that secondary recycling leads to second-grade products in most cases.

Reprocessing of recovered plastics has been another technological achievement, as in many cases special equipment has been developed. In some cases modified molding devices were chosen for compression and injection of relatively thick bodies. When relatively clean resins are obtained and pelletized, all kinds of products can be processed via conventional machinery such as extrusion into profiles, tubes and films, blow and injection molding, callendering, vacuum and rotational forming. The recycled polymers may undergo various degrees of degradation, such as chain scission (molecular weight reduction), cross-linking, oxidation and appearance of unsaturation, all of which affect chemical and mechanical properties, not to mention appearance and aesthetics (dark color, roughness and poor optical properties). (The mix of different colors may also become a drawback.) Improvement of performance can be achieved at extra cost by special additives such as impact modifiers, thermal and light stabilizers, but mainly coupling agents (so-called compatibilizers) for polyblends.

It is interesting to describe the role of recycling of various resins like PE, PP, PS, PVC, PET and engineering thermoplastics. In 1991, the total resin production in the United States was 62.8 billion pounds. Of this 14.4 billion pounds of plastics went into packaging of which 650 million pounds (4.5%) were recycled — 70% was from bottles (the majority, 50%, from PET bottles). PET has become the most successful component in recycling, due to its relatively easy collection (directly by supermarkets that returned deposits) and feasible marketing of fibers made from the cleaned and treated bottles. It can also be recycled from Mylar™ film and trays for microwave use. In addition to fibers, recovered pellets may be converted to nonedible product packaging, and some engineering appliances. The success of the other polymers, mostly commodities, is marginal. To mention PE, for example, there are LDPE, LLDPE and HDPE of various grades, so that mixing them together causes difficulties in processing (different viscosities and melting points). The LDPE recycling industry is mostly based on a collection of films and pipes from agriculture, or bags from supermarkets, as well as from dry-cleaning (garment film). When well organized this recycling scheme may become profitable, as recovered resins may be sold at 50% to 80% of the price of virgin resin. The prices rise and fall in conjunction with those of the pure resin itself. The recycled plastics may be converted to trash bags, pipes, packaging, toys and many other outlets. HDPE (either from bottles, boxes or grocery bags) may be successfully recycled and converted into bottles for nonedible liquids (detergents, oils, etc.), into pallets and other items.

PP has been recycled mainly from car batteries; however, its growth is so remarkable that it will find its own recycling path in the future, from both packaging or the automotive field. In some applications PP replaces PVC, which is less friendly to the environment. PS is not recycled much, while its use is also diminishing. The same applies to rigid PVC, while plasticized PVC may be reclaimed for the shoe and boot industry. As mentioned before, PVC is often replaced by PP, ABS and even PET (in spite of higher cost). Among the engineering plastics, PC is easily recovered directly into pallets and crates by the producers or its cooperators. PC may also serve as multiple-use bottles that can be sterilized and refilled, thus decreasing the load of waste.

Other engineering polymers (PPO, PA and POM) are recovered and reused in special construction or engineering items due to their relatively high value. Multilayer packages are more difficult to recycle, unless reworked and reused as an inner layer in new multilayer films. The thermosets are in principle nonrecyclable, but in some cases technology exists for depolymerization or pyrolysis. In principle they may be reground into fillers (incorporating fiber-reinforcement). The automotive industry has become a serious user of plastics in cars, approaching 18% in weight. They already consume 25 different types of plastics, many of which are engineering polymers and polyblends of high performance and high value. Now the industry is trying to replace many plastic parts by recyclable ones. This calls for new designs mainly for easy separation and dismantling, for reduction in the number of resins in use, and for replacing polyblends (that are more difficult for recycling) by single polymers. It is not an easy task for car engineers to comply with both high performance and recyclable elements. In this industry, as well as in other durables (machinery and household appliances), the stress of the "green" laws is enormous to cooperate in finding the best solutions.

Needless to say, metal parts must also be recycled and used cars are compacted and shredded, after which the various components are separated. This is obviously less profitable for plastic elements, but easier for steel recovery. Last but not least, what can be done with used tires that already overpack many cities in the developed countries? There is no simple solution to this problem, but various treatments, either chemical or physical, partially reduce this waste. Hydrolysis with hot caustic can break the cross-linked molecules and produce a rubbery powder, polymeric fabric, wire, and carbon black, all of which are recyclable for further use. They can be shredded and ground into crumb to replace mats. They can be used as rubber-modified sport tracks, or carpet backing. Blends of reclaimed thermoplastics and crumb rubber derived from tires are being marketed. Whole or cut tires serve in road beds, playgrounds and similar constructions. Tires can be decomposed to a fuel or incinerated for energy, although the smoke derived from carbon black is intolerable. Reclaimed tires also find some use as a substitute for new tires. In conclusion, the mountains of discarded tires (reaching about 235 million per year in the United States) are still awaiting a reasonable solution, rather than being abandoned or dumped into landfills.

If there is a conflict between plastics and ecology, we have shown that it can be overcome with imagination and talent; however, no compromise on quality and safety should be allowed. Some contribution to environmental hazards is due to the nonpolymeric components of a commercial plastic compound. The residual vinyl chloride monomer (VCM) is considered carcinogenic, so that modern regulations have limited its concentration in the polymer to ppb. Some plasticizers or solvents may also be harmful, as well as thermal stabilizers based on lead, cadmium and other heavy metal derivatives, but in most cases, appropriate substitutes have been found. The same problems exist with inorganic pigments, or fire-retardants based on halogen. So, all toxic additives should be replaced for the production of packaging or other discardable goods. To solve an environmental catastrophe related to the so-called ozone depletion (hole) that may eventually affect the whole universe, it has been found that chlorofluorocarbons (like those used as foaming agents in expanded plastics, PUR) have a significant ozone depletion potential and therefore new blowing agents are replacing the fluoro-derivatives. (Freon is also being replaced in refrigerators and cooling systems as the volatile coolant.) In summary, the coexistence of plastics and ecology is possible, when special care and regulations are used. The future of the plastics industry is tightly linked to the solution of these environmental needs.

PROBLEMS

1. What do we mean by "recyclable"?

2. What are the various options in recyclying polymers?

3. What do you know about "self-degradable polymers"?

4. Compare the overall energy recoverable from plastics to other materials for packaging.

Recommended Reading

Fried, J. R. 1995, *Polymer Science and Technology*, Prentice Hall. An advanced textbook (far beyond an introductory text), mainly for senior or graduate level of chemical engineering.

Grulke, E. A. 1994, *Polymer Process Engineering*, Prentice Hall. An extensive treatise of polymer processes, from both chemical and engineering aspects. It may serve as an advanced study after the student has been exposed to the basic concepts.

Charrier, J. M. 1991, *Polymer Materials and Processing*, Hanser. A general descriptive book on polymers and their properties and processing.

Seymour, R. B. and C. E. Carraher, 1990, *Giant Molecules*, Wiley. This is an introductory work, mainly for non-technical persons. It eliminates the engineering aspects, but provides a general preface to the world of plastics.

Seymour, R. B. and C. E. Carraher, 1988, *Polymer Chemistry*, Marcel Dekker. Covers the chemical aspects of polymer synthesis, structure and properties, but does not include polymer engineering (processing).

McCrum, M. G., C. P. Buckley, and C. B. Bucknall, 1988, *Principles of Polymer Engineering*, Oxford Science Publications. Deals with the engineering aspects of polymers, but not with their chemistry (polymerization). It provides a quantitative analysis of processing.

Rubin, I. I. (editor), 1990, *Handbook of Plastic Materials and Technology*, Wiley. An extensive review on major polymeric materials and the various technologies of fabrication. It is presented by leading authorities in polymer science and engineering.

Modern Plastics Encyclopedia, McGraw-Hill. This is the major practical encyclopedia, which appears every year and updates information on production and sales, development of new polymeric grades and technologies. Tables of valuable data of properties and processing conditions provide valuable and practical information for industry and users.

Index